Praise for *A Global Warming Primer*

Bennett's careful and question-by-question presentation will lead any fair-minded person to see the warming issue more clearly and increase understanding of the need for concern about current developments.
— **Hon. George P. Shultz**, Hoover Institution, Secretary of State under President Ronald Reagan

Jeffrey Bennett has done what many others have been unable to do: He has made climate science understandable. That is a considerable achievement given the complexity of the topic and the need for all of us to grasp the basics of what is arguably the most important topic of our time.
— **Gov. Bill Ritter, Jr.** (Colorado), author of *Powering Forward*

A Global Warming Primer delivers on its promise. In engaging, accessible, and accurate prose, Jeffrey Bennett clearly explains the science of climate change, ending with a thoughtful exploration of ways to solve the problems it poses for our future.
— **Ann Reid**, Executive Director, National Center for Science Education

This delightfully perceptive book is a must-read for everyone concerned about our future. It covers climate's complex topics in a clear, illuminating manner. The insightful approach makes the subject accessible to newcomers and brings a fresh perspective that should interest even climate experts.
— **William Gail**, President (2014), American Meteorological Society

This great book addresses common skeptical climate arguments in a way that sorts out the science from the belief on both sides of the debate. It is also bang up to date, covering the most recent analysis of the global slowdown in surface warming and changes in climate policy. I also love its optimistic focus on climate solutions.
— **Piers Forster**, IPCC Lead Author and Director, Priestley International Centre for Climate, University of Leeds, UK

A friendly yet authoritative look at how we know what we know about the climate, and why we need to do something about it.
—**Carl Zimmer**, author of *Evolution: Making Sense of Life*

A Global Warming Primer is an exceptionally valuable resource for educators at all levels. The scientific understanding of modern global warming as well as discussion of real-world solutions are made readily accessible via the book's conveniently indexed Question and Answer format coupled with an optional, deeper tier of explanation and evidence.
— **Dr. Cherilynn Morrow**, Founder of ArtSciencEducation, recipient of the American Geophysical Union SPARC Award for Education and Public Outreach

Concise, crystal clear, packed with the best available information — this is the book to grab if you want to be well informed about climate change. Carefully stepping around controversial politics, this "primer" will give you in an afternoon everything you need to know about the science and economics that will govern the future of our civilization.
— **Spencer Weart**, author of *The Discovery of Global Warming*

Those of us who are curious and concerned about climate change will not find a more lucid explanation of climate science. With clear explanations and surprisingly simple examples, Jeffrey Bennett takes the intimidation factor out of what the great majority of climate scientists have been trying to explain to us for decades.
— **William Becker**, Executive Director, Presidential Climate Action Project

A remarkably clear explanation of the causes and effects of global warming and what we can do to address it.
— **David Bookbinder and David Bailey**, Element VI Consulting

As an entrepreneur working to provide sustainable real food to communities around the world, the dangers of climate change are never far from my thoughts. For anyone who doubts the reality of the threat, this is the book to read to help you understand it. Best of all, you'll come away realizing that the problem is eminently solvable, and that the solution will help create a stronger economy and better world for our children and grandchildren.
— **Kimbal Musk**, Entrepreneur, Venture Capitalist, and Co-Founder of The Kitchen

From science to solutions, this clearly written and up-to-date survey of human-caused climate change illuminates one of the great existential issues of our time.
— **Prof. Richard C. J. Somerville**, University of California, San Diego and author of *The Forgiving Air: Understanding Environmental Change*

A creative and remarkably accessible summary of climate science and policy. Quick and easy as 1-2-3!
— **Yoram Bauman**, PhD, the "Stand-up Economist" and co-author of *The Cartoon Introduction to Climate Change*.

I recommend this book to business leaders not only to better understand the threats posed to our economy by global warming, but also to appreciate the enormous business opportunities inherent in the transition to clean energy.
— **Nicole Lederer**, Founder, Environmental Entrepreneurs (E2.org)

Author Jeffrey Bennett has an extraordinary ability to explain climate change and its impact on our planet and its inhabitants in a clear and simple manner. This is the book to read to get the true story.
— **Ben Bressler**, Founder, Natural Habitat Adventures

With clear and detailed explanations, scientist and educator Jeffrey Bennett carefully dismantles the misconceptions that have clouded the debate on climate change, then presents the solutions we must pursue to solve this critical challenge. This book is a must-read for believers and skeptics alike.
— **Andrew Chaikin**, author of *A Man on the Moon*

A book that everyone should read, whether a skeptic or a believer. Dr. Bennett clearly lays out both the arguments and the explanations, using a perfect mix of answering tough questions in a simple and straightforward way that everyone can understand, along with further background for those wanting a more detailed scientific explanation. I applaud him for taking on such a difficult but critically important topic in a way that will inspire everyone to think more hopefully about the question: What can *I* do about it?
— **Dr. Susan Lederer**, NASA Space Scientist

A very readable and understandable presentation of the basic science of global warming. This book could go a long ways in creating an informed electorate on one of the most important issues of our day.
— **Stephen Turcotte**, Professor of Physics, BYU-Idaho

A must read for everyone on our planet. *A Global Warming Primer* presents the facts while leaving the politics out in a way non-scientists can clearly understand. We hope everyone will do what he suggests and write "A Letter to Our Grandchildren" about what we as individuals will do based on the evidence.
— **Mark Levy and Helen Zentner**, Educational Consultants

A Global Warming Primer is what you get when a first-rate writer and educator brings his cosmic perspective to the most pressing issue facing humanity. Bennett's Q&A approach, while easily digestible, is rooted in complex science that few can relate so simply, clearly, and readably. Highly recommended.
— **Todd Neff**, Science Writer and author of *From Jars to Stars*

This book is just what I've been looking for as a teaching aid and primer on this subject. It elegantly summarizes a huge amount of complex information and speaks with an authoritative voice clearly based on years of experience of teaching and writing on the subject. I especially like the well-chosen quotes from conservative politicians that demonstrate that this is a challenge that overrides partisan views, and the section on hopes for the future that gives an unashamedly personal opinion without concealing the real extent of uncertainty about the options that we face. If only more commentators could give such a balanced view, we'd all be better-placed to deal with this enormous challenge.
— **James McKay**, editor of "Dreams of a Low Carbon Future" and manager of the Centre for Doctoral Training in Low Carbon Technologies, University of Leeds, UK

Make no mistake; climate change will affect all of our lives. This is understandably frightening, and we find ourselves in a time when many people prefer confusion and false controversy to facing this fear head-on. This book offers us just what we need right now: clarity. Step by step, question by question, the author states the facts, explains the underlying concepts, and offers us the best comfort we can have: the power to honestly face the facts of our changing planet. I wish everyone in the world would pause and read this book.
— **Dr. Michelle Thaller**, Astronomer, TEDx speaker

A Global Warming Primer takes a complicated topic and breaks it down in a simple way that anyone can understand while also bridging the partisan divide that sometimes gets in the way of the science. There are multiple references throughout for those who want to delve deeper into the topics, but the general format is focused and concise, making it a quick and easy read. Everyone should read this important text — our future and our children's future depend on it.
— **Gabe L. Finke**, CEO, Ascentris

Who better to help us understand global warming than the astrophysicist and educator, Jeffrey Bennett! Love how this book walks us through the scientific facts and addresses skeptic claims to provide us with intelligent talking points for discussions with families, friends, and co-workers on this important world issue.
— **Patricia Tribe**, CEO, Story Time From Space, and former Director of Education for Space Center Houston

Eschewing unnecessary and arcane details, Bennett cuts right through the noisy arguments about climate change, and shows that global warming is an inevitable consequence of simple physics and the fuels we burn. He then bears down on the essential questions: are we the victims of environmental scare-mongering? And if there's a real and present danger, what should we do? In his marvelously accessible style, Bennett tells it like it is. If you don't know what to think about climate change — or even if you do — this is the one book to read.
— **Seth Shostak**, SETI Institute and Host of *Big Picture Science*

I have read dozens and dozens of climate books and can say without equivocation that *A Global Warming Primer* should be on your short list. From science to solutions, Jeffrey Bennett provides comprehensive information in an easily understandable style.
— **Scott Mandia**, Professor of Physical Sciences, Suffolk Counting Community College

For reasons that are unreasonable to reasonable people, climate change science — and science more broadly — has been argued in recent times as a partisan subject. But this science is not blue or red; it's simply science that is essential for humankind to understand. This book admirably helps us do just that.
—**Sven Lindblad**, President & CEO of Lindblad Expeditions, Inc.

We collectively owe author Jeffrey Bennett a huge "Thank You" for this effort to enlighten anyone with questions about global warming. He has a remarkable ability to communicate complicated atmospheric and oceanic climate factors in a manner that will permit nonscientist readers to comprehend and appreciate the critical importance of this topic. Through his use of a Q&A format, he allows the reader to either dig deeply into the scientific details or take a more casual contemplation. Don't pass up the opportunity to become informed about what may well be the most important threat to our planet, both now and for the next several generations.
— **Ron Alberty**, former Chief of Meteorological Research, National Severe Storms Laboratory

I'm not a scientist, and I didn't need to be to understand this book. By presenting the evidence-based facts simply and clearly, this book will enable anyone to understand why global warming is an issue that we can't afford to ignore.
— **RJ Harrington, Jr.**, President and CEO, Sustainable Action Consulting

By sticking to the facts in an easily accessible Q & A format, Bennett deftly explores a complicated and most critical problem of our time.
— **Susan Nedell**, Rocky Mountains Advocate, Environmental Entrepreneurs (E2)

A Global Warming Primer fills a unique niche: providing a very clear and accessible description of what we know about climate change, where there are uncertainties, and the range of possible solutions. We live in a world where people's understanding of climate change is often correlated with their political beliefs — hopefully this primer will help create a fact-based understanding of the underlying science and the choices before us.
— **Dr. Will Toor**, Southwest Energy Efficiency Project, and former Mayor and County Commissioner, Boulder, CO

In clean, clear, elegant, and engaging style, Dr. Bennett lays out what scientists know about climate change and explains it in a way that will enable both young people and adults to understand what we know and how we know it. This book is a major contribution to climate literacy, taking just the right approach to engage the reader and help us all become smarter inhabitants of home planet Earth.
— **Dan Barstow**, CASIS Education Manager and Founder, Climate Literacy Network

By illustrating the challenges as well as the solutions, Jeffrey Bennett's reliance on a "big picture" approach to climate change empowers his fellow citizens to embrace a fact over fear approach to resolving our climate crisis.
— **Christina Erickson**, Attorney and Environmental Advocate

A Global Warming Primer

Also by Jeffrey Bennett

For Children
Max Goes to the Moon
Max Goes to Mars
Max Goes to Jupiter
Max Goes to the Space Station
The Wizard Who Saved the World
I, Humanity

For Grownups
Beyond UFOs: The Search for Extraterrestrial Life and Its Astonishing Implications for Our Future
Math for Life: Crucial Ideas You Didn't Learn in School
What is Relativity? An Intuitive Introduction to Einstein's Ideas, and Why They Matter
On Teaching Science: Principles and Strategies That Every Educator Should Know

High School/College Textbooks
The Cosmic Perspective
The Essential Cosmic Perspective
The Cosmic Perspective Fundamentals
Life in the Universe
Using and Understanding Mathematics: A Quantitative Reasoning Approach
Statistical Reasoning for Everyday Life

A Global Warming Primer

ANSWERING YOUR QUESTIONS ABOUT

The Science, the Consequences, and the Solutions

Jeffrey Bennett

BIG KID SCIENCE

Boulder, CO

Education, Perspective, and Inspiration for People of All Ages

Published by
Big Kid Science
Boulder, CO
www.BigKidScience.com

Education, Perspective, and Inspiration for People of All Ages

Book web site: **www.GlobalWarmingPrimer.com**

Distributed by IPG
Order online at www.ipgbook.com
or toll-free at 800-888-4741

Editing: Joan Marsh, Lynn Golbetz
Composition and design: Side By Side Studios

ISBN: 978-1-937548-78-0

To Grant and Brooke,
in hopes that you and your future children
will live your lives in a world far better than
the world of any past generation.

Brief Table of Contents

Detailed Table of Contents

Introduction

Preservation of our environment is not a liberal or conservative challenge; it's common sense.

— President Ronald Reagan, Jan. 25, 1984 (State of the Union address)

Is human-induced global warming a real threat to our future? Most people will express an opinion on this question, but relatively few can back their opinions with solid evidence. This is true on both sides, as many "believers" are no better able to explain the scientific case for global warming than "skeptics" are to make a case against it. Many times we've even heard politicians and media pundits say "I am not a scientist" to avoid the issue altogether.

But the truth is, *the basic science is not that difficult*. Sure, Earth's climate is complex, and therefore so are many details of the science of global warming, but I can tell you from my own teaching experience that the key ideas are understandable to most fourth- and fifth-graders. Indeed, even the more arcane details that you may hear debated in the media are usually simple enough once you focus on the heart of the matter. So if you want to understand and act intelligently on this issue, then I hope you'll continue reading. By the end of this book, I believe you'll be fully equipped to have an *informed* opinion about global warming.

Before we begin, let me tell you a little about the origins and goals of this book. The origins lie in my more than 30 years of experience in teaching and writing about science and mathematics. This experience has led me to conclude that one of the primary problems with science education (both in the United States and globally) is that we too often allow students — or the public — to get hung up on details before they understand the key "big picture" ideas. Global warming offers a prime example, because the media debate is filled with endless arguments about topics like the intricacies of computer models or climate feedbacks, even while the basic science is almost completely ignored. I've been fortunate to have had some success in offering a big picture approach to global warming with material I've written (along with coauthors) for my college textbooks and for a children's book (*The Wizard Who Saved the World*), as well as with my online "global warming primer." This book represents my attempt to take the same big picture approach to a wider audience.

As for goals, I'll point you to three:

1. I want to show you that anyone can understand the basic science of this issue.
2. I want to help you understand the arguments you hear from skeptics, so that you can decide for yourself what to believe.
3. I want to convince you that while the problem is real, it is also eminently solvable in ways that people of all political persuasions can agree on. Indeed, I believe that despite the "gloom and doom" you may have heard from others, the solutions not only will protect the world for our children and grandchildren but will actually lead us to a stronger economy, with energy that is cheaper, cleaner, and more abundant than the energy we use today.

As you will see, most of this book uses a question-and-answer format, with questions drawn from the many that I've been asked over the years. I hope that this format will make the book feel at least a little more like a personal discussion. You'll also notice that the book uses two distinct font sizes. The normal font is for general text (like this introduction) and the big picture ideas that should be of interest to all readers, while this smaller font is for more detailed discussion that you can treat as optional, depending on the depth to which you'd like to go.

It is my fervent hope that this book will help you see through the fog of the media debate and the partisan battles to the simple truth embodied in the quote on the first page from Ronald Reagan: When it comes to preserving the environment in which we all live, we just need a little common sense.

The Basic Science — Easy as 1-2-3

What we are now doing to the world . . . by adding greenhouse gases to the air at an unprecedented rate . . . is new in the experience of the Earth. It is mankind and his activities which are changing the environment of our planet in damaging and dangerous ways.

— British Prime Minister Margaret Thatcher, Nov. 8, 1989 (speech to the United Nations)

We all know that human activities are changing the atmosphere in unexpected and in unprecedented ways.

— President George H. W. Bush, Feb. 5, 1990 (remarks to the Intergovernmental Panel on Climate Change)

The two quotes above show that, more than a quarter century ago, the conservative leaders of both the United Kingdom and the United States were already convinced of the reality and the threat of global warming. What made them so sure? In the case of Thatcher, it probably helped that she was a scientist herself (trained in chemistry), which made it easier for her to recognize the underlying scientific ideas. But Bush was not a scientist, and he and many other people of all political persuasions were still able to understand the same ideas. Why? Because they are not very difficult.

In this first chapter, I'll show you the simple underlying science of global warming. Note that nothing in this chapter is subject to any scientific debate at all, and you'll find that this basic science is accepted even by the scientists who count themselves as ardent skeptics (as evidence, look ahead to the quote that opens chapter 2). To show you just how simple and solid it is, let's begin with an example from astronomy.

A Tale of Two Planets

Figure 1.1 shows the planets Earth and Venus to scale, along with their global average surface temperatures. You can see that both planets are about the same size; they also both have about the same overall composition of rock and metal. But look at the enormous difference in their surface temperatures.[1] Earth has temperatures ideally suited to life and our civilization, while Venus is hot enough to melt lead. If you think about it, you might wonder why two planets that are so similar in size and composition would have such drastically different surface temperatures.

It might be tempting to chalk it up solely to the fact that Venus is closer to the Sun than Earth, but that is not the answer. Figure 1.2 shows part of the Voyage Scale Model Solar System, which shows the sizes and distances of the Sun and planets on a scale of 1 to 10 billion. Notice that while it's true that Venus is closer to the Sun, the difference isn't really all that great, and it's not nearly enough to account for such a large temperature difference by itself. Moreover, Venus's bright clouds reflect so much sunlight that its surface actually absorbs *less* sunlight than Earth's, which by itself would lead us to expect Venus to be *colder* than Earth. So why is Venus so hot?

The primary answer is carbon dioxide, a gas that can trap heat and make a planet warmer than it would be otherwise. In fact, as we'll discuss in more detail shortly, both planets would actually be frozen over if they had no carbon dioxide in their atmospheres at all. Earth has just enough carbon dioxide (plus water vapor; see page 13) to make our planet livable, so in that sense, carbon dioxide is a very good thing for

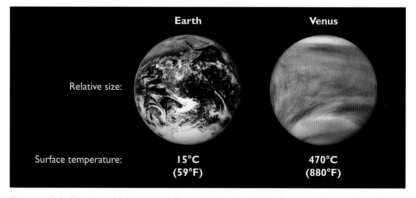

Figure 1.1 Earth and Venus are shown to scale. Why do two planets that are so similar in size and composition have such different surface temperatures?

---

1 In science, temperatures are almost always stated on the Celsius (°C) scale, which is also the temperature system used in most of the world. In this book, I will generally also provide Fahrenheit (°F) equivalents, since those are more familiar in the United States.

Figure 1.2 This photo shows the inner portion of the Voyage Scale Model Solar System, located outside the National Air and Space Museum (Washington, D.C.). The locations of the Sun, Venus, and Earth are indicated. The model Sun is the visible gold sphere; on this scale, Venus and Earth are each about the size of the ball point in a pen (about one millimeter in diameter) and can be seen within the glass disks that face outward from their pedestals.

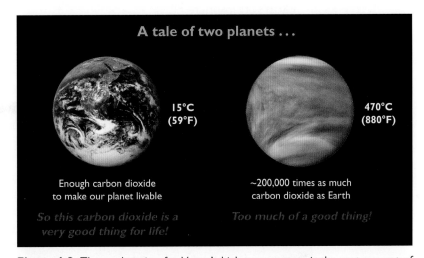

Figure 1.3 The explanation for Venus's high temperature is the vast amount of carbon dioxide in its atmosphere, which traps an enormous amount of heat through what we call the *greenhouse effect*.

life. But Venus has almost 200,000 times as much carbon dioxide in its atmosphere as Earth, and all this carbon dioxide traps so much heat that the entire surface is baked hotter than a pizza oven — providing clear proof that it is possible to have too much of a good thing (figure 1.3).

This story of Venus and Earth contains almost everything you need to understand the basic science of global warming. It shows that gases like carbon dioxide, which we call *greenhouse gases*, really do make planets warmer than they would be otherwise, and that the more of these gases a planet has, the hotter it will be.

Global Warming 1-2-3

The lesson from our tale of two planets leads directly to the subtitle of this chapter, in which I say that global warming is as easy as 1-2-3. By this, I mean that for all the arguments you may hear in the media, the basic science of global warming can be summarized in three simple statements, which embody two indisputable scientific facts and the inevitable conclusion that follows from them:

1. ***Fact:*** Carbon dioxide is a greenhouse gas, by which we mean a gas that traps heat and makes a planet (like Earth or Venus) warmer than it would be otherwise.
2. ***Fact:*** Human activity, especially the use of fossil fuels[2] — by which we mean coal, oil, and gas, all of which release carbon dioxide when burned — is adding significantly more of this heat-trapping gas to Earth's atmosphere.
3. ***Inevitable Conclusion:*** We should *expect* the rising carbon dioxide concentration to warm our planet, with the warming becoming more severe as we add more carbon dioxide.

Notice the inevitability of the conclusion: As long as both of the facts are true — and I'll show you why there is no scientific doubt about either of them — then there's really no way around the conclusion that global warming should be expected.

Of course, knowing that global warming is expected doesn't tell us how badly or imminently we'll be affected, and by itself it leaves open the possibility that other factors (such as climate feedbacks) might mitigate or even counteract the expected warming, at least on some time scales. We'll discuss the debate over these issues in chapter 2. First, however, we'll turn our attention to the evidence that supports our two facts.

---

2 Fossil fuels (coal, oil, gas) get their name from the fact that they come from the remains (fossils) of living organisms that died and decomposed long ago. They are rich in carbon because all life on Earth is based on carbon. When they burn, their carbon combines with oxygen to produce carbon dioxide.

Evidence for Fact 1 (Carbon Dioxide Makes Planets Warmer)

Fact 1 is that carbon dioxide is a greenhouse gas that makes a planet warmer than it would be otherwise. Now, in Q&A format, we're ready to examine the evidence that makes this a fact rather than a matter of opinion.

How do we know that Fact 1 is really a fact?

There is no doubt that higher concentrations of carbon dioxide and other greenhouse gases make planets warmer, because this fact is based on the simple, well-understood, and well-tested physics of what we call the *greenhouse effect*. Figure 1.4 shows how the greenhouse effect works. Notice the following key ideas:

- The energy that warms Earth comes from sunlight, and in particular from visible light (the kind of light that our eyes can see). Some sunlight is reflected back to space, and the rest is absorbed by the surface (land and oceans).
- Earth must ultimately return the energy it absorbs back to space. The returned energy takes the form of *infrared* light, which our eyes cannot see.
- Greenhouse gases — which include water vapor (H_2O), carbon dioxide (CO_2), and methane (CH_4, also commonly called *natural gas*) — are made up of molecules[3] that are particularly good at absorbing infrared light. Each time a greenhouse gas molecule absorbs a *photon* (the technical name for a "piece" of light) of infrared light, it quickly reemits it as another infrared photon, which may head off in any random direction. This photon can then be absorbed by another greenhouse gas molecule, which does the same thing.

The net result is that greenhouse gases tend to slow the escape of infrared light from the lower atmosphere, while their molecular motions heat the surrounding air. In this way, the greenhouse effect makes the surface and the lower atmosphere warmer than they would be from sunlight alone. The more greenhouse gases present, the greater the degree of surface warming. A blanket offers a good analogy. You stay warmer under a blanket not because the blanket itself provides

---

3 A brief review in case you've forgotten from elementary school science: All ordinary matter is made up of atoms, but sometimes atoms are bound together in *molecules*. We state molecular composition with simple formulas, like H_2O, which means a molecule made up of two hydrogen (H) atoms and one oxygen (O) atom. Similarly, carbon dioxide is CO_2 because it consists of molecules with one carbon (C) atom and two oxygen (O) atoms, while methane is CH_4 because it has one carbon (C) atom and four hydrogen (H) atoms.

The Greenhouse Effect

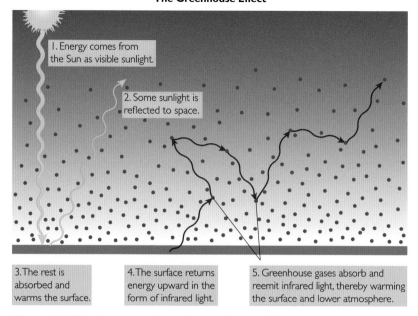

1. Energy comes from the Sun as visible sunlight.

2. Some sunlight is reflected to space.

3. The rest is absorbed and warms the surface.

4. The surface returns energy upward in the form of infrared light.

5. Greenhouse gases absorb and reemit infrared light, thereby warming the surface and lower atmosphere.

Figure 1.4 This diagram explains the greenhouse effect, which makes a planet's surface and lower atmosphere warmer than they would be otherwise. The yellow arrows represent visible light, the red arrows represent infrared light, and the blue dots represent greenhouse gas molecules.

any heat, but because it slows the escape of your body heat into the cold outside air.

Q **How do we know that Earth returns energy to space in the form of infrared light?**

It's basic physics, verified by observations. All objects — including the Sun, the planets, and even you — always emit some form of light,[4] but the form depends on the temperature. Hot objects, like the Sun, emit visible light. Cooler objects, like planets and you, emit only infrared light. While we cannot see infrared light with our eyes, we can detect it with infrared cameras and other instruments, and orbiting satellites have directly measured the amount of infrared light being emitted by Earth.

Q **Why haven't you mentioned nitrogen and oxygen, which make up most of our atmosphere?**

The atmosphere is indeed made mostly of nitrogen and oxygen; together, these two gases make up about 98% of the gas in Earth's atmosphere (77% for nitrogen and 21% for oxygen). However, molecules of nitrogen and

---

4 More technically, objects generally emit what we call *thermal radiation* (also called *blackbody radiation*) that has a characteristic spectrum in which the emitted light has an intensity and range of wavelengths that depend only on the object's temperature. Hotter objects have spectra that peak at shorter wavelengths, while also emitting more energy per unit area at all wavelengths, than cooler objects.

oxygen do not absorb infrared light, and therefore do not contribute to the heating of the surface. In other words, without the relatively small amounts of infrared-absorbing greenhouse gases (such as water vapor, carbon dioxide, and methane) that are present in our atmosphere, all the infrared light emitted from Earth's surface would escape directly into space, and our planet would be frozen over.

In case you are wondering why some molecules can absorb infrared light and others cannot, it is a result of their structures. In our atmosphere, nitrogen and oxygen both take the form of molecules in which two atoms are bound together; that is, nitrogen is in the form N_2 and oxygen in the form O_2. In order to absorb photons of infrared light, molecules must be able to vibrate and rotate. This turns out to be fairly difficult for molecules with only two atoms, particularly when both atoms are the same, as in N_2 and O_2; that is why these molecules do not contribute to planetary heating. In contrast, vibration and rotation are relatively easy for many molecules with more than two atoms, which is why water vapor (H_2O), carbon dioxide (CO_2), and methane (CH_4) all absorb infrared light effectively, making them greenhouse gases.

Are there any other greenhouse gases I should know about?

Although water vapor, carbon dioxide, and methane are the three most important greenhouse gases in Earth's atmosphere, other trace gases can also act as greenhouse gases, which means they can also contribute to warming. Those that you are likely to hear about and that we'll discuss a bit more in this book include nitrous oxide (N_2O) and industrial chemicals known as halocarbons, which include chlorofluorocarbons (CFCs).

I've heard that "greenhouse effect" is a misnomer. Is that true?

It depends on how picky you want to be. The term comes from botanical greenhouses, but those greenhouses actually trap heat through a different mechanism than planetary atmospheres: Rather than absorbing infrared radiation, greenhouses stay warm primarily by preventing warm air from rising. Nevertheless, atmospheric greenhouse gases and botanical greenhouses have the same net effect of keeping things warmer than they would be otherwise, so I'm personally fine with the term "greenhouse effect."

How do we know that greenhouse gases really trap heat?

Two major lines of evidence show conclusively that greenhouse gases trap heat. First, scientists can measure the heat-trapping effects of these gases in the laboratory. Although the actual setups are somewhat more complex, the basic idea is simply to put a gas (such as carbon dioxide) in a tube, shine infrared light at it, and measure how much of that light passes through and how much is absorbed. Such measurements were first made more than 150 years ago by British scientist John Tyndall (figure 1.5) and have been repeated and refined ever since.

Second, we can easily confirm that the greenhouse effect raises actual planetary temperatures in the way we discussed earlier for Earth and Venus. If there were no greenhouse effect, a planet's average

John Tyndall's Setup to Measure Light (1859)

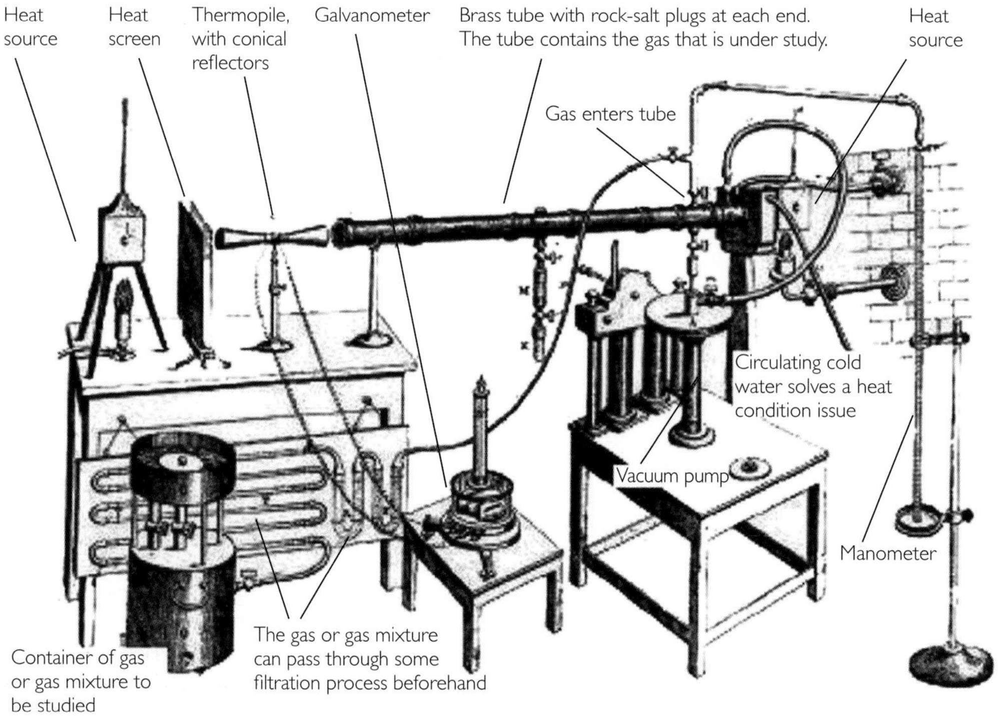

Heat source | Heat screen | Thermopile, with conical reflectors | Galvanometer | Brass tube with rock-salt plugs at each end. The tube contains the gas that is under study. | Heat source

Gas enters tube

Circulating cold water solves a heat condition issue

Vacuum pump

Manometer

Container of gas or gas mixture to be studied

The gas or gas mixture can pass through some filtration process beforehand

Figure 1.5 This diagram shows the experimental setup used by John Tyndall in 1859, when he first measured how gases like carbon dioxide create what we now call the greenhouse effect. The measurements have been repeated and refined ever since. Source: The original illustration is from Tyndall's 1872 book *Contributions to Molecular Physics in the Domain of Radiant Heat*; this annotated version is from Wikipedia.

temperature would depend only on its distance from the Sun and the relative proportions of sunlight that it absorbs and reflects. I won't bother you with the mathematical details, but they lead to the simple formula that you can see being applied to Earth in figure 1.6. The formula shows that Earth's global average temperature would be well below freezing (–16°C, or +3°F) without greenhouse gases. In other words, we need the greenhouse effect to explain Earth's actual average temperature, which is about 15°C (59°F). The same is true for all other planets: We get correct answers for planetary temperatures only when we use mathematical formulas that include the greenhouse effect.

This brings us back to our tale of two planets. For Earth, we find that without the naturally occurring greenhouse effect, our planet would be too cold for liquid oceans and life as we know it. That is why, as we saw in figure 1.3, the natural greenhouse effect is a very good thing for life on Earth. But Venus, with almost 200,000 times as much carbon dioxide in its atmosphere as Earth, clearly has much too much of this good thing.

$$T_{\text{no greenhouse}} = 280°C \sqrt{\frac{(1-\text{reflectivity})}{d^2(AU)}} - 273°C$$

$$\text{Earth:} \quad = 280°C \times \sqrt{\frac{(1-0.29)}{1^2}} - 273° = -16°C$$

Earth actual $= 15°C$

$\Rightarrow$ greenhouse warming $= 31°C$

Figure 1.6 This painting shows the calculation of Earth's expected average temperature if there were no greenhouse effect. The fact that this temperature (−16°C) is so much lower than the actual average temperature (+15°C) shows that the natural greenhouse effect is what makes Earth warm enough for life. Painting by Roberta Collier-Morales from *The Wizard Who Saved the World*.

Q Why does Venus have so much carbon dioxide in its atmosphere?

Earth actually has about the same total amount of carbon dioxide as Venus, but while Venus's carbon dioxide is virtually all in its atmosphere, nearly all of Earth's is "locked up" in what we call *carbonate rocks*, the most familiar of which is limestone. The reason for this difference is that Earth has oceans and Venus does not.

On both planets, the original source of carbon dioxide was gas released by volcanoes. On Earth, carbon dioxide dissolves in the oceans (which contain about 60 times as much carbon dioxide as the atmosphere), where it then combines with dissolved minerals to form carbonate rocks (which contain almost 200,000 times as much carbon dioxide as the atmosphere). Venus lacks oceans and therefore cannot dissolve carbon dioxide gas, so it all remains in the atmosphere.

A deeper question is why Earth has oceans and Venus does not, and scientists attribute this to the fact that Venus is closer to the Sun (Venus is about two-thirds as far from the Sun as Earth). You can understand the role of distance from the Sun by thinking about what would happen if Earth

The Basic Science

If Earth moved to Venus's orbit

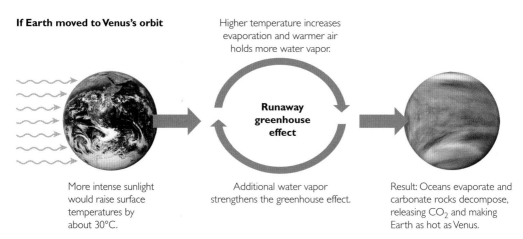

Higher temperature increases evaporation and warmer air holds more water vapor.

Runaway greenhouse effect

More intense sunlight would raise surface temperatures by about 30°C.

Additional water vapor strengthens the greenhouse effect.

Result: Oceans evaporate and carbonate rocks decompose, releasing CO_2 and making Earth as hot as Venus.

Figure 1.7 This diagram shows what would happen if Earth magically moved to Venus's orbit.

were magically moved to Venus's orbit (figure 1.7). The greater intensity of sunlight would immediately raise Earth's average temperature from its current 15°C to about 45°C (113°F). The higher temperature would increase the evaporation of water from the oceans, putting much more water vapor into the atmosphere — and because water vapor is a greenhouse gas, the added water vapor would strengthen the greenhouse effect and drive temperatures even higher. The higher temperatures, in turn, would lead to even more ocean evaporation and more water vapor in the atmosphere, strengthening the greenhouse effect even further. In other words, we'd have a *reinforcing feedback*[5] process in which each little bit of additional water vapor in the atmosphere would lead to a higher temperature and even more water vapor. The process would rapidly spin out of control, resulting in what scientists call a *runaway greenhouse effect*. It would not stop until the "moved Earth" became as hot as (or even hotter than) Venus is today.

In fact, something like this probably occurred on Venus long ago. Based on scientific understanding of how the Sun generates energy through nuclear fusion, the Sun should very gradually brighten with time. The rate is so slow that we cannot measure it, but the calculations indicate that the Sun was about 30% dimmer when the planets were born (about 4½ billion years ago) than it is today. This means that the young Venus probably had sunlight of not much greater intensity than Earth does today, and some scientists suspect that Venus may have had oceans at that time. As the Sun gradually brightened, Venus grew hotter until a runaway greenhouse effect set in.

Q Does Mars also have a greenhouse effect?

Yes, but it is very weak. The atmosphere of Mars is made mostly of carbon dioxide (about 95%), but the atmosphere is so thin (the surface pressure is less than 1% of that on Earth) that the total amount of carbon dioxide is actually quite small. As a result, Mars is warmed only a little by its greenhouse effect, and its greater distance from the Sun makes it quite cold, with

---

5 In science, a reinforcing feedback is more commonly called a *positive feedback*; it is a feedback that adds to and thereby amplifies an existing effect, in contrast to a *negative feedback*, which reduces an existing effect.

an average surface temperature of –50°C (–58°F). Scientifically, the surprise is that Mars shows clear evidence of having had liquid water on its surface in the past. This means that long ago, Mars must have been much warmer than it is today, which in turn means that it must once have had a much stronger greenhouse effect. Scientists have a pretty good idea of why Mars once had a strong greenhouse effect and why the effect ultimately weakened so much, but the full discussion isn't directly relevant to our topic in this book.[6]

Why are you focusing on carbon dioxide, when there's more water vapor (also a greenhouse gas) in the atmosphere?

It's true that there is more water vapor than carbon dioxide in the atmosphere. In fact, there's about 10 times as much water vapor as carbon dioxide, and water vapor does indeed contribute more than carbon dioxide to Earth's overall greenhouse warming. However, *carbon dioxide is the more critical gas in setting Earth's temperature.*

The reason is that once we increase the carbon dioxide concentration of the atmosphere, the concentration tends to remain higher for many decades, centuries, and even millennia. In contrast, water vapor cycles easily into the atmosphere through evaporation and out of the atmosphere through rain and snow. As a result, the amount of water vapor in the atmosphere at any given time is determined by the temperatures of the ocean and atmosphere. That is, the amount of water vapor in the atmosphere can change in response to temperature changes, but it does not initially cause those changes. Instead — and very importantly — water vapor *amplifies* climate changes initiated by other factors, because it acts as a reinforcing feedback. For example, if more carbon dioxide raises the global temperature a little bit, the atmosphere can hold more water vapor, which then traps more heat, making the temperature rise even more. Conversely, if the carbon dioxide level drops, the global temperature decreases so that there is less water vapor in the atmosphere, which then traps less heat so that the temperature drops further. This amplification by water vapor is well understood and is necessary to explaining Earth's natural cycles of ice ages and warm periods, which we'll discuss in chapter 2.

Can you be more precise about how long added carbon dioxide remains in the atmosphere?

Yes, but we also have to phrase the question more precisely. Let's pose it this way: If we suddenly stopped adding carbon dioxide to the atmosphere, how long would it take for the carbon dioxide concentration to drop back down to something closer to its "natural" (preindustrial) value? To answer this

---

6 You can find many sources for learning more about our current scientific understanding of Mars and how its climate changed, including discussions in my astronomy and astrobiology textbooks (*The Cosmic Perspective* and *Life in the Universe*).

question, scientists do calculations based on all the various ways in which carbon dioxide can be removed from the atmosphere, which include uptake by plants, dissolving in the oceans, and the gradual production of seashells and carbonate rocks. The details are fairly complex and subject to some uncertainties, and the answer also depends on how much carbon dioxide we've added (in total) by the time we stop adding it, but here's a brief summary of current understanding:

For the first few decades, uptake by the land and oceans would remove carbon dioxide relatively rapidly, so that about a third of the carbon dioxide we'd added would be removed in 20 to 50 years. But then the rate would slow dramatically: Between 15% and 40% of our added carbon dioxide would still remain in the atmosphere after 2,000 years, and it would take tens of thousands of years for the concentration to come all the way back down to its preindustrial value.[7] (This discussion assumes natural processes only, as opposed to "geoengineering" schemes in which we develop technologies that can remove carbon dioxide from the atmosphere.)

What about other greenhouse gases?

After carbon dioxide and water vapor, methane (CH_4) is the next most abundant greenhouse gas in Earth's atmosphere, and though it is much less abundant than carbon dioxide and makes a much smaller overall impact on the climate, its effect is still important. The same is true for other greenhouse gases, and scientists can and do take all these greenhouse gases into account when gauging the strength of the greenhouse effect. However, aside from a brief discussion that you'll find on pages 22–23, I will generally ignore these other gases in this book, primarily because carbon dioxide has the greatest impact on Earth's climate and we can therefore keep our science discussions simpler by keeping the focus on it. Nevertheless, the contribution of other greenhouse gases is very important to consider both scientifically and in policy decisions, so while they may be secondary in importance to carbon dioxide, they should not be ignored.

What's the bottom line for Fact 1?

There is no doubt at all about the fact that atmospheric carbon dioxide and other greenhouse gases create a *greenhouse effect* that makes a planet warmer than it would be otherwise. The heat-trapping effects of greenhouse gases have been measured in the laboratory, and our overall understanding of the greenhouse effect has been further verified by the fact that actual planetary temperatures match calculated temperatures only when we take it into account. Indeed, while a small minority of scientists dispute the *threat* of global warming, you will not find any legitimate scientists who dispute the basic physics of the greenhouse effect.

---

7 A good summary of these ideas is presented in the IPCC Working Group 1 Report (2013), Chapter 6, FAQ 6.2, which can be downloaded from www.ipcc.ch/report/ar5/wg1/.

Evidence for Fact 2 (Human Activity Is Adding Carbon Dioxide to the Atmosphere)

We now turn to the evidence that establishes Fact 2, which is that human activity, especially the use of fossil fuels, is adding heat-trapping carbon dioxide to Earth's atmosphere.

How do we know that the atmospheric concentration of carbon dioxide is really rising?

The most direct way to measure the amount of carbon dioxide in the atmosphere is to collect and study air samples. Scientists have been making such direct measurements continuously since the late 1950s. Figure 1.8 shows the measurements made using samples collected at the Mauna Loa Observatory in Hawaii. As you can see, the measurements clearly show a rapidly rising concentration of carbon dioxide in Earth's atmosphere. Measurements made at numerous other sites around the world show similar increases over time.

Notice that the units on the vertical axis are *parts per million* (ppm), which means the number of carbon dioxide molecules among each 1 million total molecules of air. You can see that the carbon dioxide concentration has recently surpassed 400 parts per million, which is the

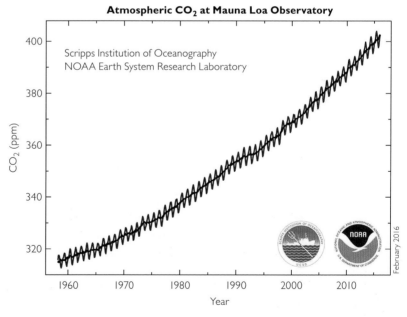

Figure 1.8 This graph shows direct measurements of the amount of carbon dioxide in the atmosphere, which have been made on a regular basis since the late 1950s. Source: National Oceanographic and Atmospheric Administration (NOAA). The data are updated monthly, and you can always see the latest at www.esrl.noaa.gov/gmd/ccgg/trends.

same as 0.04%.[8] In other words, carbon dioxide represents only a tiny fraction of the molecules in Earth's atmosphere. Nevertheless, as we've already discussed, this small amount of carbon dioxide is very important because of its role in the greenhouse effect.

Q What are the small wiggles on the graph?

These small wiggles represent seasonal variations in the carbon dioxide concentration. They occur because plants and trees absorb carbon dioxide as they grow in spring and summer, then release it as they decay in fall and winter. The global pattern follows the seasons of the Northern Hemisphere because, as you'll see if you look at a globe, that is where most of Earth's land mass — which also means most of the plants and trees — is located. (The Southern Hemisphere is mostly ocean.) The seasonal wiggles peak each year in May because that is when most of the prior year's vegetation has decayed (releasing its carbon dioxide), while the Northern Hemisphere's summer growing season is not yet far enough along for new vegetation to have absorbed very much carbon dioxide.

Q Are measurements at Mauna Loa really representative of the whole world?

The carbon dioxide concentration varies somewhat from place to place on Earth, so it's important to choose measurement locations that are relatively unaffected by local conditions and therefore representative of changes occurring in the atmosphere as a whole. The Mauna Loa site was selected by Scripps Institution scientist Charles David Keeling because its high altitude and its location on a relatively isolated island make the air above it representative of a large portion of the global atmosphere. Today, scientists also measure the carbon dioxide concentration at many other locations around Earth, and these measurements confirm that the concentration is rising at a similar rate around the world. Note also that measurements are made by several independent scientific groups (for example, scientists from Scripps and from the National Oceanic and Atmospheric Administration [NOAA]), and these different measurement sets agree very well, further confirming that the measured changes are real. We usually show the Mauna Loa data because they constitute the longest continuous record from any site and because other data confirm that they are representative of the global carbon dioxide concentration. Incidentally, Keeling's work has proven to be so important that the graph shown in figure 1.8 is often called the *Keeling curve* in his honor.

Q Can we measure the carbon dioxide concentration further into the past?

Although we only have direct measurements since the 1950s, scientists have discovered a variety of ways to measure the carbon dioxide concentration from earlier times. The most reliable records come from air bubbles trapped in ancient ice — that is, ice that has remained frozen for

---

8 Here's how to see this: 400 parts per million means a concentration of $400 \div 1{,}000{,}000$, which equals $4/10{,}000$. Percentages are essentially parts per 100, and $4/10{,}000$ is the same as $0.04/100$, or 0.04%.

Figure 1.9 These photos show the drilling of an ice core, which consists of the compressed snows (with trapped air bubbles) laid down year after year. Scientists can read the past climate history in the layers in much the same way that we can learn about the past from tree rings. Source: NASA, Goddard Space Flight Center; photos copyright by photographer Reto Stöckli; reprinted with permission.

long periods of time in glaciers or in Greenland or Antarctica. Although the work is difficult and requires great care, the basic idea is simple: Scientists drill down into the ice and bring up an *ice core* (figure 1.9). The ice core is made from the accumulated snow of many years, which over time is compressed into solid ice. The deeper layers represent ice laid down at earlier times. The longest ice core ever drilled extended to a depth of about 3.2 kilometers (2 miles) in the Antarctic ice, and represents snows that fell and accumulated over a period of 800,000 years.

By studying the air bubbles in this long ice core, scientists have been able to measure how the carbon dioxide concentration has varied in Earth's atmosphere over the past 800,000 years. Figure 1.10 shows the results, along with a zoom-out to show the direct measurements made since the 1950s. Notice that the carbon dioxide concentration has risen and fallen substantially many times over the 800,000-year period. These variations must be natural, because they predate the human

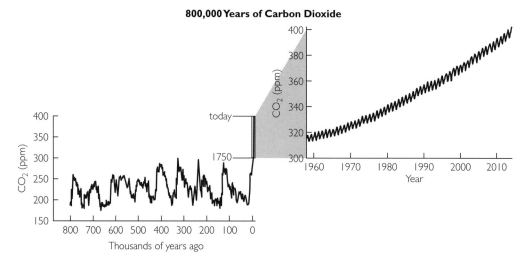

Figure 1.10 This graph shows how the atmospheric carbon dioxide concentration has changed over the past 800,000 years. Source: Data from the European Project for Ice Coring in Antarctica. Also see the animated version of this graphic at www.esrl.noaa.gov/gmd/ccgg/trends/history.html.

burning of fossil fuels, which became important only in the past couple hundred years.

Several key facts should jump out at you as you study figure 1.10:

- The carbon dioxide concentration has varied naturally over the past 800,000 years, but only within a range between about 180 and 290 parts per million. These natural variations are dwarfed by the huge increase that has occurred since the industrial revolution began in about 1750.

- Today's carbon dioxide concentration of about 400 parts per million is already about 40% higher than the pre-industrial concentration (280 parts per million in 1750), which was itself near the highest found at any time in the past 800,000 years.

- If we extrapolate into the future, we find that the carbon dioxide concentration is on track to reach *double* its preindustrial value (560 parts per million) in just 50 to 60 more years, and *triple* that value (840 parts per million) by the middle of the next century.

How do we know the added carbon dioxide is a result of human activity, rather than natural sources?

It's true that there are natural sources that can add carbon dioxide to the atmosphere (such as volcanoes), but we can be very sure that the recent, dramatic rise in carbon dioxide that you see in figure 1.10 is due almost entirely to human activity. Most of this carbon dioxide comes from the burning of fossil fuels, with lesser amounts from

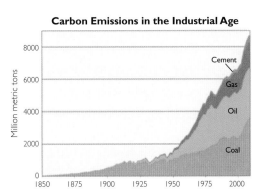

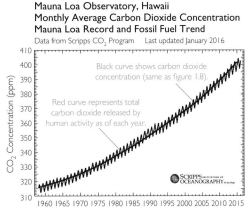

Figure 1.11 (Left) This graph shows the amounts and sources of carbon dioxide released by human activity since 1850. (Right) This graph shows that the rate of increase in the total amount of carbon dioxide released by human activity (red curve) tracks almost perfectly with the measured rise in the atmospheric carbon dioxide concentration (black curve). Sources: (left) Reproduced directly from J. M. Melillo, T. C. Richmond, and G. W. Yohe, eds., 2014, *Highlights of Climate Change Impacts in the United States: The Third National Climate Assessment* (U.S. Global Change Research Program); (right) Scripps CO2 Program (scrippsco2.ucsd.edu/ graphics_gallery/mauna_loa_record/mauna_loa_fossil_fuel_trend).

deforestation and industrial processes, such as cement production,[9] that release carbon dioxide. There are four major reasons why we can be so sure that humans are responsible for the current rise in the carbon dioxide concentration.

First, the rise in atmospheric carbon dioxide coincides almost perfectly with the increased release of carbon dioxide by human activity. The ice core data tell us that for the 1,000 years prior to 1750, the atmospheric carbon dioxide concentration stayed very close to 280 parts per million; the dramatic rise that has taken it past 400 parts per million today began right when the industrial revolution began, which is also when humans first began to use large quantities of fossil fuels. This correlation becomes even clearer when you look at how the release of carbon dioxide due to industry tracks with the increase in the carbon dioxide concentration. The graph on the left side of figure 1.11 shows how the amount of carbon dioxide released by human activity each year has been rising with time, and identifies the amounts from different sources. Now look at the graph on the right side of figure 1.11. At first, you may think you are just looking at a repeat of figure 1.8 with different colors. However, in figure 1.8, the black curve

---

9 Cement production involves the heating of carbonate minerals (such as limestone), which then release some of their stored carbon dioxide. Cement currently contributes about 5% of the carbon dioxide that we are adding to the atmosphere. Numerous companies are working on new cement production methods that offer the promise of reducing or eliminating these carbon dioxide emissions.

running up the middle was the average trend running through the seasonal wiggles. This time, the seasonal wiggles are shown in black, and the red curve running up the center represents the rising amount of carbon dioxide that has been released into the atmosphere by humans.[10] The virtually perfect tracking shows that the atmospheric carbon dioxide concentration is rising precisely as we'd expect if its source is human activity.

Q Does *all* the carbon dioxide released by human activity add to the atmospheric concentration?

No. Careful measurements show that only about half of the carbon dioxide released by humans each year is staying in the atmosphere. Much of the rest is being dissolved into the oceans (and some is taken up by plants and soil on land). Measurements confirm that the carbon dioxide concentration is increasing in the oceans in tandem with the increase in the atmosphere. This increase is making ocean water slightly more acidic, which creates the problem of *ocean acidification* that we will discuss in Chapter 3 as one of the major consequences of global warming.

Worth noting: Although we do not know exactly how much carbon dioxide the oceans are capable of absorbing, scientists expect the rate of uptake to slow as the oceans warm. If it does, then as the oceans absorb less, more of the carbon dioxide released by human activity will stay in the atmosphere, which will increase the rate at which the carbon dioxide concentration is growing even beyond that shown in figure 1.8.

Second, there really isn't any other possibility. Scientists have a variety of ways to measure the amount of carbon dioxide that is added by natural sources, such as volcanoes, and it just doesn't compare to the amounts being released by the burning of fossil fuels and other human activity. In fact, the natural contributions are smaller than about 1% of the human contributions.

Q Wait — I heard that the amount of carbon dioxide released by human activity is very small compared to the amount released by natural sources like the oceans and living organisms. So why do you say that the natural sources aren't contributing to the increase?

It's because the natural sources are in a natural balance. That is, while it's true that the amount of carbon dioxide released from the oceans and by life is far larger than the amount released by human activity, those natural releases are almost perfectly balanced by natural processes that absorb carbon dioxide. For example, not counting the extra added by human activity,

---

10 To be a little more precise, the red curve (in the right graph of figure 1.11) represents the total (cumulative) emissions over time. In other words, you could construct this curve by taking the year-by-year data from the graph on the left side of figure 1.11 and adding them together to get the total amount released from 1850 up to any particular year. The units for the red curve are not shown, because the important fact is the match in the *rate* of rise of the two curves. For more details on the graph, see the Web site of the Scripps CO_2 Program.

the oceans always absorb essentially the exact same amount of carbon dioxide that they release, and plants naturally absorb all the carbon dioxide exhaled by animals (and people). You can be sure this is the case because if it weren't, the carbon dioxide concentration would always vary wildly over a huge range, rather than staying within the fairly narrow range that you see in figure 1.8. The only way that natural processes can change the carbon dioxide concentration is by being out of balance. Volcanoes can disrupt the balance, since their eruptions add carbon dioxide without removing it, but as noted above, these amounts are small compared to the human release. Deforestation also adds carbon dioxide, because it releases carbon dioxide stored in trees and plants — but this isn't exactly a natural process, since humans are the cause of most of the deforestation that has occurred in the past few centuries.

Third, the burning of fossil fuels consumes oxygen at the same time that it releases carbon dioxide, which means that if the rising carbon dioxide concentration comes from fossil fuels, there should be a corresponding *decrease* in the concentration of oxygen in the atmosphere and oceans — and this has indeed been measured. The oxygen decrease probably won't have direct effects on us, because oxygen makes up about 21% of Earth's atmosphere and to date the decline has lowered this value by less than 0.1%. However, it may be contributing to an increase in low-oxygen areas of the ocean, which can be very damaging to ocean life.

Fourth, and perhaps most convincing of all, careful chemical analysis of atmospheric carbon dioxide shows changes that only make sense with fossil fuels as the source. The key to understanding this evidence is to recognize that carbon atoms come in three different forms, or *isotopes*, known as carbon-12, carbon-13, and carbon-14, and the relative abundances of these three isotopes are different in carbon that comes from different sources (such as volcanoes, deforestation, and the burning of fossil fuels). Therefore, we can determine the source of atmospheric carbon dioxide by measuring these isotope abundances. Let's first consider carbon-14, which is radioactive and exists today only as a result of ongoing production as cosmic rays hit atoms in Earth's upper atmosphere.[11] Carbon-14 becomes incorporated into living organisms through respiration, but it decays after the organisms die, and the organisms that made fossil fuels died so long ago that there is no carbon-14 in fossil fuels at all. As a result, if fossil fuels are the source of the rising carbon dioxide concentration, then the relative abundance of atmospheric carbon-14 (compared to ordinary carbon-12) should be

---

11 Carbon-14 has a half-life of about 5,700 years, which means that any carbon-14 that existed when Earth first formed is long gone. The fact that carbon-14 decays with a known half-life in the remains of living organisms that have died is what makes "radiocarbon dating" possible for fossils and archeological artifacts up to a few tens of thousands of years old. (For older fossils, the carbon-14 will have all decayed, so scientists date them with radioactive isotopes that have longer half-lives.)

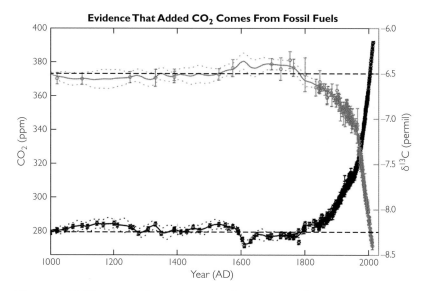

Figure 1.12 This graph shows the atmospheric carbon dioxide concentration (black) and the relative abundance of carbon-13 (brown) over the past 1,000 years, as measured from ice cores. The declining abundance of carbon-13 is a "smoking gun" that leaves no doubt that the rising carbon dioxide concentration is coming mostly from the burning of fossil fuels. (The carbon-13 abundance is shown as "δ13," which is a standard scientific unit used for this purpose.) Source: M. Rubino et al., *J. Geophys. Res. Atmos.* 118 (2013): 8482–8499, doi:10.1002/jgrd.50668.

falling as the total carbon dioxide rises — and this is just what has been observed.

Even more impressive isotope evidence comes from changes in the relative abundance of carbon-13. Overall, carbon-13 represents about 1.07% of all natural carbon on Earth, but its percentage is slightly lower in living organisms (because life incorporates carbon-12 more readily into living tissues than carbon-13), which means it is also slightly lower in fossil fuels (since they are the remains of living organisms). Figure 1.12 shows ice core data for the past 1,000 years with the total carbon dioxide concentration in black and the relative abundance of carbon-13 in brown. Notice that the carbon-13 abundance has been dropping in tandem with the rise in carbon dioxide, just as we should expect if the rising carbon dioxide comes from the burning of fossil fuels (with their lower abundance of carbon-13). In effect, these isotopic data are a "smoking gun" that leaves no doubt that most of the added carbon dioxide is coming from the burning of fossil fuels.

Is human activity also increasing the concentrations of other greenhouse gases?

Yes. The left graph in figure 1.13 shows changes in the atmospheric methane concentration since the late 1970s (the time period for which

Other Greenhouse Gases

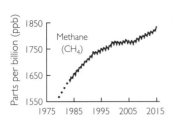 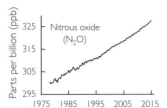 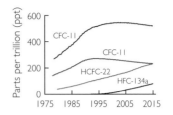

Figure 1.13 These graphs show the concentrations measured since the late 1970s of methane, nitrous oxide, and halocarbons, including CFCs and gases used as substitutes for CFCs. Source: NOAA Earth System Research Laboratory, www.esrl.noaa.gov/gmd/aggi/.

direct measurements of all the gases shown in the figure are available). Data from ice cores and other sources show that the methane concentration has more than doubled since 1750. Human activity adds methane to the atmosphere in several ways, but the three largest are (1) agriculture, in which methane is released from rice paddies and from the raising of livestock; (2) oil and gas extraction and transport, during which methane can leak directly into the atmosphere; and (3) landfills, in which decomposing wastes release methane.

The middle graph in figure 1.13 shows the rapidly rising nitrous oxide concentration. Nitrous oxide is released primarily through the production and use of fertilizers, which means these emissions are tied to food production. It's worth noting that because the release of nitrous oxide is unrelated to the use of fossil fuels, it's essentially a separate problem that we will need to deal with along with carbon dioxide and methane emissions.

The rightmost graph in figure 1.13 shows halocarbons, which come entirely from human manufacturing and do not exist naturally. Notice that while CFC concentrations were rising rapidly in the 1970s and 1980s, their concentrations have since declined. The reason for the decline is that, starting in the 1970s, scientists began to recognize that CFCs could cause destruction of Earth's atmospheric ozone layer, which protects us from dangerous ultraviolet light from the Sun. As a result, the nations of the world came together to sign the global treaty known as the Montreal Protocol (and subsequent revisions to strengthen it), which has successfully led to a dramatic decline in the production and use of CFCs.

To gauge the combined effects of carbon dioxide and other greenhouse gases, scientists use something called the "annual greenhouse gas index" (AGGI), which you can learn more about from the AGGI section of the NOAA Web site. These measurements indicate that the combined greenhouse warming from *all* gases emitted through human activity is roughly 50% larger than that from carbon dioxide alone.

What's the bottom line for Fact 2?

There is no doubt that human activity is adding carbon dioxide and other greenhouse gases to the atmosphere. We know the carbon dioxide is from human activity because the rate of increase is what we expect based on the rate at which humans are releasing carbon dioxide; because natural sources are too small to account for the observed increase; because we see a corresponding decrease in atmospheric oxygen, just as expected; and because chemical analysis shows that the added carbon dioxide is coming from the burning of fossil fuels. Moreover, as Margaret Thatcher said in the quote that opens the chapter, we are adding this carbon dioxide at an "unprecedented rate" that is "new in the experience of the Earth," at least for the past 800,000 years (and probably since our planet was born).

Global Warming 1-2-3: The Inevitable Conclusion

Let's repeat the "global warming 1-2-3" that we started out with:

1. **Fact:** Carbon dioxide is a greenhouse gas, by which we mean a gas that traps heat and makes a planet (like Earth or Venus) warmer than it would be otherwise.

2. **Fact:** Human activity, especially the use of fossil fuels — by which we mean coal, oil, and gas, all of which release carbon dioxide when burned — is adding significantly more of this heat-trapping gas to Earth's atmosphere.

3. **Inevitable Conclusion:** We should *expect* the rising carbon dioxide concentration to warm our planet, with the warming becoming more severe as we add more carbon dioxide.

In this chapter, I've shown you the evidence that explains why Facts 1 and 2 are scientifically supported beyond any reasonable doubt. Because statement 3 — that we should *expect* global warming to be occurring — follows inevitably, you can probably see where concern for global warming comes from.

Of course, knowing we should expect global warming does not tell us how fast it will occur or how detrimental its consequences will be. To understand those issues, we need to investigate the climate in more detail, and we'll do that in the next chapter. But if we keep adding carbon dioxide to the atmosphere, our planet *will* become warmer. You might therefore begin asking yourself, how much warming are you willing to risk?

2   The Skeptic Debate

I will simply try to clarify what the debate over climate change is really about. It most certainly is not about whether climate is changing: it always is. It is not about whether CO_2 is increasing: it clearly is. It is not about whether the increase in CO_2, by itself, will lead to some warming: it should. The debate is simply over the matter of how much warming the increase in CO_2 can lead to, and the connection of such warming to the innumerable claimed catastrophes.

— **Richard Lindzen, Feb. 22, 2012 (speech to the British House of Commons)**

Richard Lindzen is arguably the most prominent "skeptic" disputing the threat of global warming, primarily because he has strong scientific credentials as a professor at MIT. He has been called to testify before Congress many times, he has had numerous articles published in media outlets that argue against global warming concerns (such as the *Wall Street Journal* editorial pages), and he speaks frequently to groups that oppose action on global warming. And yet, as you can see from his quote above, even he does not dispute the basic scientific case that we discussed in chapter 1. He disputes only the magnitude of the threat.

In other words, while there is no doubt that global warming is real, there *is* some legitimate debate about how serious a problem it is. This is where you may have heard about the "scientific consensus" on global warming — namely, that the vast majority of scientists who have studied the issue have concluded that it is a serious threat to our future that therefore demands serious and immediate action. Still, it's only fair to acknowledge that it's *possible* that Lindzen and other skeptics could be right when they claim that the threat of global warming has been overblown. In this chapter, I'll explain the four major points of debate that have been raised by the skeptics, and we'll examine what the evidence says in each case.

Skeptic Claim 1: Earth Is Not Warming Up as Expected

The first skeptic claim we sometimes hear about global warming is that despite what I've told you about the simplicity of the 1-2-3 science, Earth is not actually warming up the way that logic predicts. To test this claim, we simply need to look at the evidence.

Is the world actually warming up?

To determine whether Earth is actually warming up, scientists need to track changes in Earth's global average temperature over time. Direct measurements from which we can infer global temperature go back to about 1880, though there are greater uncertainties for the earlier years. Figure 2.1 shows the data. Notice the clear upward trend — for an overall gain of at least about 0.85°C, or 1.5°F, over the past century — which confirms that our world is indeed warming, just as our 1-2-3 logic told us to expect.

Are these temperature data reliable?

Yes, but only thanks to some very careful work. Measuring Earth's global average temperature essentially requires scientists to average local temperature measurements from many places around Earth, and this is not easy to do. For example, three fairly obvious complexities are: (1) even today, there are large regions of our planet (including the oceans and regions near the poles) for which we have relatively few temperature measurements, making it difficult to come up with a fully global average; (2) this problem becomes worse as we look to the past, when there were fewer weather stations; and (3) many measurements are made in or near urban areas, which tend to produce higher temperatures than they would if the same regions were rural or unpopulated, because urban areas generate their own heat through such means as the absorption of sunlight by pavement and the heat emitted from cars and homes (the "urban heat island effect").

Because of these and other difficulties, there's always some uncertainty in Earth's precise global average temperature. In fact, the estimate of 15°C that I gave earlier (see figure 1.3) could be off by as much as a degree or two. That is why figure 2.1 shows only temperature *differences* (scientists often call them "anomalies") from year to year, rather than actual values. To understand how this helps, imagine weighing yourself every day on two different scales, one of which always gives you a lower weight than the other. You may not have any way to know which scale is showing your true weight, but if you actually lose five pounds in a week, both scales will probably show the same five-pound loss. In much the same way, year-to-year differences measured by weather stations are much more reliable than their exact temperature readings. Therefore, by averaging year-to-year differences measured at weather stations around the world, scientists can get a reliable record of how Earth's temperature is changing, even without knowing the "true" average temperature. Moreover, for recent decades,

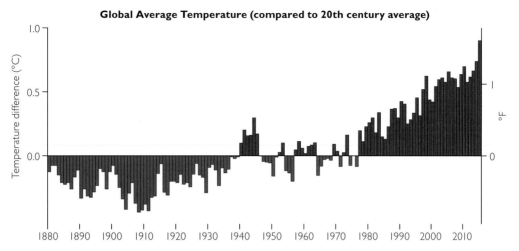

Figure 2.1 This graph shows how global average temperature varied from 1880 through 2015. The horizontal line (for 0°) represents the average temperature for the entire 20th century. Notice the clear warming trend of recent decades. Source: National Climate Data Center (NOAA).

scientists also have data from satellites,[1] which in effect can take measurements from all around the world, including the regions where no weather stations are located.

That said, it's still not easy. For example, the numbers and locations of weather stations change over time, the heat in cities can change as they grow, and different satellites collect data in different ways. Scientists must be very careful to take these factors into account when computing the change in temperature from one year to the next. Fortunately, several different scientific groups analyze both ground and satellite temperature data, each using somewhat different techniques.[2] The results found by these different groups are all in close agreement, giving scientists great confidence that the trend shown in figure 2.1 is real. Indeed, while there is some debate over the size of the uncertainties in the data, there is no serious controversy over the general trends, which show that the world has been getting warmer over the past century.

Finally, it's worth noting that the warming trend shown in figure 2.1 probably *underestimates* the true change. The reason is that polar regions are underrepresented in the data (because they have relatively few weather stations), and the available data show clearly that these regions are warming more than others. Therefore, if we had as many weather stations in polar regions as we do in other places, the data would probably show even greater warming.

---

1   However, satellites cannot directly measure temperature at the surface, and instead give readings for temperatures at roughly the altitudes at which airplanes fly (8–15 kilometers), where warming is less pronounced than at the surface. For this reason, satellite temperature measurements must be interpreted with great care.
2   If you want to learn more about these groups and how they measure the global average temperature, a good starting point is this Web page: www.carbonbrief.org/blog/2015/01/explainer-how-do-scientists-measure -global-temperature/.

Q How much uncertainty is there in the data in figure 2.1?

Scientists generally state measurement uncertainty (often referred to as a "margin of error") in terms of some level of confidence, and the most common level used is "95% confidence" (which you may hear scientists refer to as "two sigma"). For example, if a measurement is stated as "0.8°C with an uncertainty of 0.1°C" (also written as 0.8°C ± 0.1°C), it means that there is a 95% chance that the true value is between 0.7°C and 0.9°C. Note that these uncertainty statements are *not* mere guesswork; they are based on careful analysis of the data and the potential sources of error in the measurements. Of course, careful analysis is hard, and for this reason different groups may not always agree on either the "best value" of a measurement or the precise uncertainty range. Still, as I've already noted, different groups analyzing the temperature data have found results that are all in close agreement. As you might expect, the uncertainties are greater for times further in the past (when there were fewer weather stations and no satellite measurements). Overall they are approximately as follows:[3]

- For the early years in figure 2.1 (e.g., 1880–1900), the uncertainty in the measurements (with 95% confidence) is about 0.1°C. For example, the bar for 1885 shows –0.2°C, so the true value (with 95% confidence) was likely between –0.3°C and –0.1°C.
- The uncertainty becomes smaller as time goes on, and for recent decades (since about 1980) is down to about 0.03°C. For example, the bar for 2015 shows a value of 0.90°C, so the true value (with 95% confidence) was likely between 0.87°C and 0.93°C.
- For the overall warming trend of 0.85°C since 1880, the uncertainty is about 0.2°C, so the total warming (with 95% confidence) has probably been between 0.65°C and 1.05°C.

Q **Wait — I've heard that global warming has stopped since the late 1990s. Is the world still warming up?**

Perhaps the favorite claim of the skeptics in recent years has been that global warming has "stopped" (or "paused") since the late 1990s. But this claim is demonstrably false. We expect temperatures and the climate system to have some natural variability, and this is apparent if you look at the year-to-year changes. Therefore, if you want to understand long-term trends, you have to look at averages over periods of multiple years. In figure 2.2, I've replaced the year-by-year data from figure 2.1 with the *average* (mean) for each five-year period. Notice that while there has been some *slowing* of the upward trend since the late 1990s, the trend remains upward. Therefore, *global warming has not stopped*. Indeed, as you can see, every five-year period since 1980 has set a new record for the hottest (since 1880). The most recent five-year

3 In fact, most scientists find the uncertainties to be even smaller than those I have given here, but I prefer to err on the side of being more conservative. You can find further discussion of the uncertainties at the Web site for the GISS (Goddard Institute for Space Studies) Surface Temperature Analysis. And if you feel you understand statistics well enough to read a detailed analysis of the measurements and uncertainties, see J. Hansen et al., "Global Surface Temperature Change," *Rev. Geophys.* 48, RG4004 (2010).

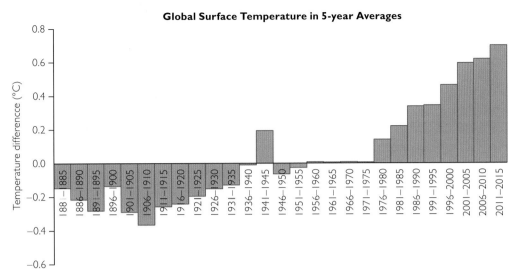

Figure 2.2 This graph shows the temperature data from figure 2.1 grouped into five-year averages. Notice that while there was some slowing of the rate of increase in the past 15 years, every five-year period since 1980 has set a new record.

period (2011–2015) is no exception, and if you look back at figure 2.1, you'll also see that 2014 and 2015 set back-to-back one-year records, with 2015 shattering the prior record by a large margin. We *don't* expect every coming year to set a new record, but the evidence clearly shows an ongoing warming trend, and the basic science tells us to expect this general trend to continue in the future.

Q Still, shouldn't there be some explanation for the slowing?

Yes, there should be, and while scientists are still trying to understand the details, the basic explanation almost certainly goes as follows. The additional heat and energy trapped in the atmosphere by the rising carbon dioxide concentration can manifest itself in several different ways, and the rising surface temperature shown in figures 2.1 and 2.2 is only one of those. In fact, more than 90% of the added heat and energy is expected to warm the water *in* the oceans (as opposed to warming the land and ocean surface), and data indicate that the ocean waters have continued to warm without any evidence of slowing (figure 2.3). Indeed, a recent study (P. J. Gleckler et al., *Nature Climate Change* [Jan. 18, 2016]) indicates that the warming of the ocean water *accelerated* during the same period in which the surface warming slowed. Another area where the additional heat and energy can show up is in glacial melting, and there is similarly no sign of a slowdown in this melting. In other words, the most likely explanation for the slowing of the temperature increase since the late 1990s is simply that more of the added heat was deposited to the oceans and glacial melting during this period than during other periods.

If you are wondering why the heat would be deposited in different ways at different times, this is actually expected as a result of natural factors in

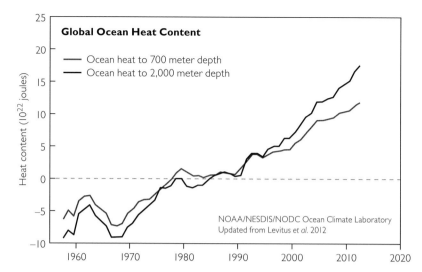

Figure 2.3 This graph shows how the measured heat content of the oceans has changed in recent decades; the data are plotted as five-year moving averages. Notice that there has been no slowing of the rise in ocean heat content, and in recent years more of the heat has been appearing in deeper waters. Source: NOAA, www.nodc.noaa.gov/OC5/3M_HEAT_CONTENT/.

the climate. For example, you've probably heard of the weather phenomenon known as "El Niño," which has numerous effects but is most noticeable as a warming of the eastern Pacific Ocean. El Niño events occur naturally and typically last about a year, but their precise length can vary significantly, and they recur at irregular intervals and at varying strengths. Because El Niño events affect the entire Earth, they can change the way heat is deposited.[4] The same is true of other natural processes in Earth's climate.

It's also worth noting that if you look carefully at figure 2.1, you'll see that 1998 was an exceptionally warm year compared to the years on either side of it. Indeed, if you were to remove 1998 from the data set, the "slowing" of the temperature rise since the 1990s would be much less pronounced. Why was 1998 so warm? It was a strong El Niño year; in fact, it came during the strongest El Niño in decades — though a similarly strong one is under way as I write in early 2016, and may explain why 2015 broke the previous one-year temperature record by such a large margin.

Bottom line: Global warming has not stopped, and while the rate of increase in the surface temperature has slowed since the late 1990s, this slowing is probably a temporary phenomenon in which more heat has been going into the oceans and glacial melting, leaving less heat going into the atmosphere. We can therefore anticipate that the upward trend in surface temperatures will continue, and if the heat absorption by the oceans slows, then the surface temperature increase will likely accelerate.

---

4 To learn more about El Niño, I recommend the brief discussion at www.climate.gov/news-features/blogs/enso/what-el-niño–southern-oscillation-enso-nutshell and the more detailed analysis, including a graph showing the strength of different El Niño events, at www.climate.gov/news-features/understanding-climate/climate-variability-oceanic-niño-index.

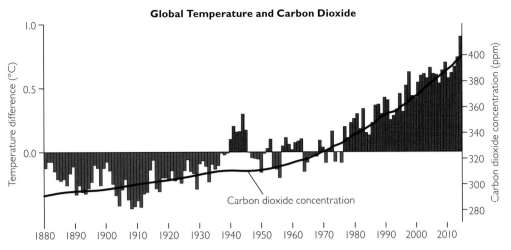

Figure 2.4 This graph repeats the temperature data from figure 2.1, with an overlay showing the carbon dioxide concentration (as an average for each year, so as to avoid seeing the seasonal wiggles shown in figure 1.8). The two are clearly moving in tandem for recent decades, lending support to the simplicity of our 1-2-3 logic for global warming.

Q **I recently read of a data reanalysis suggesting that there was no slowing. What's up with that?**

A 2015 paper published in *Science* magazine (T. R. Karl et al., *Science* 348, no. 6242 [June 26, 2015]: 1469–1472) has suggested that the actual rise in temperatures has been greater for recent years than that shown in figure 2.1. As I write this in early 2016, scientists are still debating whether this new claim is correct. I cannot claim any particular expertise on this issue, but based on discussions I've had with colleagues, I think that most scientists still assume the slow down was real. Either way, note that this claim would not in any way undercut the evidence of warming; if it is correct, it means the warming has been even *greater* in recent years than what I have shown you already.

Q **Does the warming match up with what we might expect from the carbon dioxide rise?**

Yes. Our 1-2-3 logic suggests that the observed warming and the rising carbon dioxide concentration ought to be moving hand in hand, at least in a general sense. Figure 2.4 shows that this is indeed the case.

Q **So is there any way that Skeptic Claim 1 could be correct?**

A couple of decades ago, there were still enough uncertainties in the temperature measurements that some scientists wondered if the warming trend was real. For that reason, a great deal of effort was put into understanding the uncertainties, and while some still exist (as we've discussed), there is no longer any serious debate about the trend.

2

The Skeptic Debate

In fact, the only people who still question the general warming are those, including a few prominent media pundits and politicians, who claim that the entire issue of global warming is some kind of hoax. But the evidence we've discussed is the product of work by thousands of scientists who have dedicated their lives to obtaining accurate and reliable data, and who have carefully examined both the strengths and the weaknesses of the data. So unless you believe that these thousands of scientists from around the world are all coordinating some great conspiracy, there's no reasonable doubt about the fact that our world is warming.

Skeptic Claim 2: It's Warming Up, but It's Natural

As we've discussed, there is no longer any serious scientific debate about the general warming trend. However, a few skeptics — including some with scientific training — have suggested that the warming may be occurring for natural reasons, rather than as a result of human activity. So let's look at the evidence to see whether there is any possibility that natural factors rather than human activities are the cause of the observed warming trend.

Could the Sun be the cause of the observed global warming?

The Sun does indeed vary in its energy output from year to year, though by a very small amount (much less than 1%), which means small changes in the amount of sunlight reaching Earth over time. Moreover, we know that even relatively small changes in the amount of sunlight reaching Earth can affect the climate; as we'll discuss shortly, such changes have probably been the triggers for cycles of past ice ages. But we can be very confident that changes in sunlight are *not* the cause of recent global warming, because of the data shown in figure 2.5. This figure compares changes in Earth's temperature since 1880 (red curve) to changes in the amount of sunlight reaching Earth (blue curve). Notice that while the two trends matched up moderately well until about 1950, they have since gone in opposite directions. Clearly, we cannot blame an increase in temperature on a decrease in sunlight.

Q. Can you explain the curves in figure 2.5 more clearly, including why they use an 11-year average?

Notice that each curve is actually two curves: a solid one showing the 11-year average and a ghosted one showing year-to-year data. Let's start

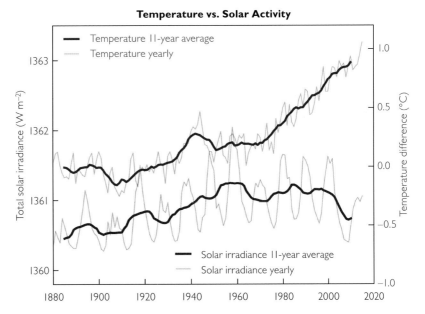

Figure 2.5 This graph compares changes in global average temperature since 1880 (essentially the same data shown in figure 2.1) to the amount of sunlight reaching Earth over the same period of time. Notice that, for recent decades, the amount of sunlight has moved in the opposite direction of the observed warming, which means the Sun cannot be the cause of recent global warming. *Note:* As you can see on the graph, the technical term for the amount of sunlight reaching Earth is *solar irradiance*, and it is measured in units of watts per square meter. Source: SkepticalScience.com.

with the red temperature curves. The ghosted red curve represents the same data shown in figure 2.1, but as a line graph rather than a bar chart. The solid curve is what we call a *moving average* (or "running mean") drawn through the first curve. That is, instead of showing temperatures for individual years, which vary quite a bit, each value on the solid curve represents the average for several years around it. This particular red curve shows each value as the average over 11 years, which means five years before and five years after each year for which it is plotted. (The solid curve stops before the end of the data set because we don't yet have a full five years of "after" data for the most recent five years.)

The blue curves are similar. The ghosted curve shows actual year-to-year data, while the solid curve shows an 11-year moving average. One subtlety: For recent decades, the data on the amount of sunlight reaching Earth are based on actual measurements made by orbiting satellites. Earlier data are reconstructed based on historical observations of sunspots, which have been reliably recorded since long before the satellite era. The sunspot observations can be translated into solar irradiance because sunspot numbers correlate very well with the amount of sunlight.

That last subtlety explains why the graph uses 11-year averages: As you can see if you look closely at the ghosted blue curve, the number of sunspots on the Sun varies in an approximately 11-year cycle. Therefore, an 11-year moving average is the "fairest" way to show the data, because it effectively removes the variations due to the sunspot cycle so that we can see the underlying general trend.

That is already pretty definitive, but there's also a second reason we can rule out the Sun as the cause of the recent warming. If the Sun were responsible for global warming, we would expect the extra sunlight reaching Earth to warm the surface and the entire atmosphere more or less uniformly. In contrast, while the greenhouse effect warms Earth's surface and lower atmosphere, it actually *cools* Earth's upper atmosphere (that is, in the *stratosphere* and above).[5] Observations show that the upper atmosphere is cooling, just as expected with a strengthening greenhouse effect, and the opposite of what we'd expect if global warming were being caused by the Sun.

In fact, several additional patterns of warming are also consistent with a strengthening greenhouse effect but not consistent with changes in the Sun.[6] For example, only greenhouse warming can account for measurements showing that nights have warmed more than days and winters (in both hemispheres) have warmed more than summers. Moreover, satellite measurements show that the total heat radiating into space from Earth has declined at the specific wavelengths radiated by carbon dioxide, which can only mean that this heat is being trapped by carbon dioxide molecules through the greenhouse effect.

Could it be other natural factors besides the Sun?

As we've just discussed, the pattern of warming is fully consistent with its being due to the addition of greenhouse gases through human activity. Still, Earth's climate is very complex and affected by many factors, both human and natural, so it's worth exploring whether there might be any other natural process that can explain the observed warming. The primary way that scientists investigate this possibility is by using what we call *models* of the climate.

Scientific models differ from the models you may be familiar with in everyday life, which are typically miniature representations of real objects, such as model cars or airplanes. In contrast, a scientific model

---

5   The precise reasons why the greenhouse effect leads to upper atmospheric cooling are fairly complex and beyond the scope of this book, but they are based on detailed calculations of how the greenhouse effect works. If you want evidence that these calculations are valid, just look to Venus, where the extremely strong greenhouse effect causes not only the very high surface temperature that we've already discussed, but also upper atmospheric cooling that matches the predictions of greenhouse calculations.
6   For a more complete discussion of the "fingerprints" that indicate the warming is from the greenhouse effect and not the Sun or other natural factors, see www.skepticalscience.com/its-not-us.htm.

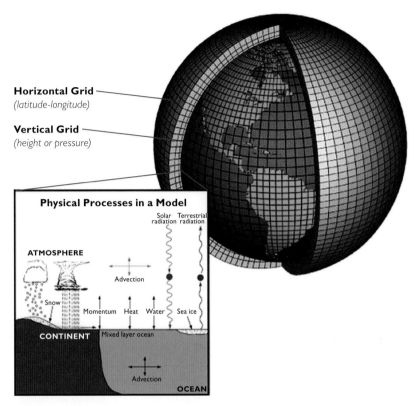

Horizontal Grid
(latitude-longitude)

Vertical Grid
(height or pressure)

Physical Processes in a Model

Solar Terrestrial
radiation radiation

ATMOSPHERE

Advection

Snow

Momentum Heat Water Sea ice

CONTINENT Mixed layer ocean

Advection

OCEAN

Figure 2.6 This illustration summarizes how a climate model works. A computer program represents Earth's climate in a series of cubes. In each cube, scientists input data from some point in time to represent "initial conditions," then "run" the model by using equations that represent the physical processes that can change the initial conditions. Source: NOAA.

is a conceptual representation, often developed with the help of computers, that uses known scientific laws, logic, and mathematics in an attempt to describe how some aspect of nature works. The model can be tested by seeing how well it corresponds to reality. Models are important in almost every field of science, but here we'll focus specifically on models of Earth's climate.

The principle behind a climate model is relatively simple. Scientists create a computer program that represents the climate as a grid of cubes like those shown in figure 2.6, so that each cube represents one small part of our planet over one range of altitudes in the atmosphere. The "initial conditions" for the model consist of a mathematical representation of the weather or climate within each cube at some moment in time. This representation might incorporate data on such things as the temperature, air pressure, wind speed and direction, and humidity at the time the model begins. The model uses equations of physics (for example, equations that describe how heat flows from one cube to neighboring cubes) to predict how the conditions in each cube will change in some time period, such as the next hour. It then uses the new conditions and the equations to predict the conditions after another hour, and so on. In this way, the model can simulate climate changes over any period of time.

Decades ago, climate models were fairly simple, using grids no more complex than the one in figure 2.6. Over time, however, scientists have in essence used trial and error to make the models better and better. Again, the principle is easy to understand: If your model fails to reproduce the real climate in some important way, then you look to see what might be going wrong. For example, you might have neglected some important law of physics, or the cubes in your grid might need to be smaller to give accurate results. Once you think you know what went wrong, you revise the model, and see if it works better. If it does, then you have at least some reason to think you are on the right track, and if it doesn't, you go back to the drawing board.

Today's climate models are fantastically detailed, and they reproduce the actual climate of the past century with remarkable accuracy. Indeed, the modern models work so well that scientists can use them to conduct "experiments" in which they ask what would happen if this or that were different than it is. Figure 2.7 shows an example of the power this approach provides. The red curve shows temperatures over the past century and a half as predicted by the best available climate models, which take into account both natural factors affecting climate, such as changes in the Sun's output and volcanic eruptions, and human factors, such as the increase in the carbon dioxide concentration from the burning of fossil fuels. Notice that these models provide an excellent match to the general trends in the real data (black curve). In contrast, models that leave out the human factors predict the blue curve, and as you can see, this curve does not agree with the observed warming of the past few decades. The fact that we get a close match between the models and reality only when changes in both natural *and* human factors are included gives us great confidence that human factors are the cause of the recent warming.

Q Why are you saying "models" in plural?

It is not possible to create an exact representation of Earth's climate (because it is too complex), so approximations must inevitably be used. Over the past few decades, numerous research groups around the world have made decisions about these approximations and developed their own climate models, each of them unique. While this might at first sound like a mess, it actually makes our confidence in modeling stronger, because despite their differences, all of these models now yield very similar results. We'd only expect this to be the case if all the models are successfully taking into account the most important climate factors. The model

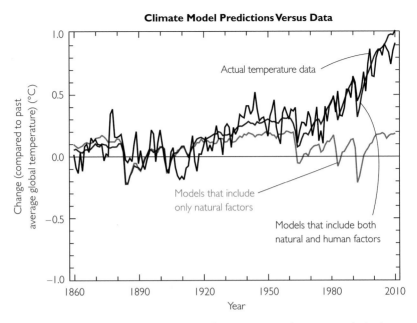

Figure 2.7 This graph shows the excellent agreement between today's climate models (red curve) and actual temperature changes (black curve), and the clear failure of models (blue curve) that take into account only natural factors in the climate. Conclusion: Today's climate models work extremely well and demonstrate that global warming is caused by human factors such as the rising carbon dioxide concentration. Source: Intergovernmental Panel on Climate Change (IPCC). *Note:* Bloomberg Business created an outstanding set of graphics to show how the natural factors combine to make the blue curve, which you can see at www.bloomberg.com/graphics/2015-whats-warming-the-world/.

curves in figure 2.7 represent averages of results from several different individual models.

What's the bottom line for Skeptic Claim 2?

There are no known natural factors that could account for the substantial warming of the past century. We've discussed two sets of observations that definitively rule out the Sun as the cause: (1) solar energy input has been falling while the temperature has been rising; and (2) the upper atmosphere has been cooling while the lower atmosphere warms, which is consistent only with greenhouse warming, not warming due to the Sun. Scientists investigate other potential causes with models, and today's sophisticated models match up extremely well with observations of the actual climate — but only when we include the human contributions to global warming, not natural factors alone. The match makes it highly likely that the models are on the right track, giving us further confidence in the idea that human activity is the cause of most or all recent global warming.

Skeptic Claim 3: It's Warming Up, Humans Are Causing It, but It's Nothing to Worry About

The evidence for human causation of climate change is now so strong that very few skeptics still dispute the idea of human-caused global warming. Instead, as Lindzen's quote at the beginning of this chapter indicates, the more common skeptic claim is that the scientific consensus overestimates the level of danger posed by the warming. This claim tends to come in three major forms, each of which we'll investigate here:

1. Skeptics point out that the climate has varied naturally in the past, and we are still here.
2. Skeptics claim the future warming will be less than most models predict.
3. A few skeptics suggest that warming may even be beneficial, rather than something to concern us.

Skeptic Claim 3, Part 1: Natural Climate Variability

There is no question that Earth's climate varies naturally over time, and skeptics have seized on this fact in two major ways. Some have used it to argue that the current warming might simply be part of a natural cycle, but we've already discussed the fact that natural factors seem unable to explain this warming. A more legitimate debate is over whether the current warming is a danger, given what we know about past climate change. So let's investigate.

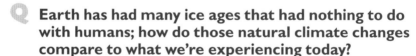 **Earth has had many ice ages that had nothing to do with humans; how do those natural climate changes compare to what we're experiencing today?**

Earth has indeed cycled in and out of ice ages in the past, and we obviously did not cause any of these changes. We can study changes in Earth's average temperature over the past 800,000 years with the very same ice cores used to measure past carbon dioxide concentrations (see figures 1.9 and 1.10). In brief, careful study of the layers in the ice cores allows scientists to make fairly precise estimates of the temperatures at the times the layers were laid down.[7]

---

7 More specifically, the temperature information is derived from careful measurements of isotope ratios (particularly of oxygen-18 to oxygen-16 and deuterium to hydrogen), cross-checked against other available data. The details are beyond the scope of this book, but if you want to read about them, a good starting point is this NASA Web page about ice core measurements: earthobservatory.nasa.gov/Features/Paleoclimatology_IceCores/.

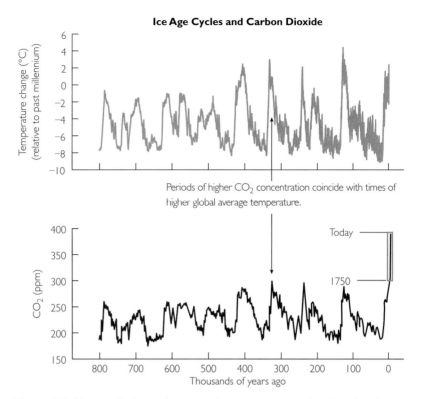

Figure 2.8 This graph shows changes in the temperature and carbon dioxide concentration over the past 800,000 years. Notice the close correlation, just as we expect from our basic 1-2-3 science. Source: Data from the European Project for Ice Coring in Antarctica.

Figure 2.8 shows the temperature record from the ice cores, along with the carbon dioxide record that we saw previously in figure 1.10. Notice that temperatures have fluctuated significantly over the past 800,000 years. The cool periods are ice ages, and the warm periods (known as "interglacials") come in between them. Moreover, just as we expect from the basic science, the warm periods match up with higher carbon dioxide concentrations and the cool periods with lower carbon dioxide concentrations. This is further confirmation that our basic 1-2-3 science really is correct.

The skeptics point to these natural changes to suggest that the changes we are causing today are nothing to worry about. But consider these key points:

- The current temperature is already approaching the highest it has been in the past 800,000 years. Given that the current carbon dioxide concentration is some 40% higher than at any other time in that period — and rising rapidly — it would seem that we should be very concerned about how much higher the temperature will rise.

• Although the figure makes it look like the onset of warm and cool periods occurred fairly rapidly, when you consider that it shows 800,000 years, you'll realize that "fairly rapidly" still means "over centuries." In contrast, the changes we are causing today are happening over decades. Again returning to the chapter 1 opening quote by Margaret Thatcher, these changes are "new in the experience of the Earth."

To summarize, while it's true that the climate changes naturally, today we are causing changes of an unnatural degree at an unnatural rate. It's hard to see how anyone could take any comfort from these facts.

Q What causes the natural cycles of ice ages and warm periods?

The observed pattern of ice ages and warm periods lines up very well with a pattern of small, cyclical changes in Earth's axis tilt and orbit that arise from gravitational effects of the Sun, Moon, and planets on Earth. These cyclical changes are called *Milankovitch cycles*, after a Serbian scientist who investigated their role in climate change. (Search on the name to learn more about these cycles.) But there's a very important point that goes along with this: By themselves, the changes that would occur as a result of the Milankovitch cycles are *not enough* to explain the large temperature swings that occur. Instead, these cycles are "triggers" that initiate feedback processes that amplify the temperature changes.

Here's how it is thought to work when a warm period begins: The changes due to the Milankovitch cycles slightly increase the amount of sunlight warming Earth and the oceans. This warming causes the oceans to release some of their dissolved carbon dioxide into the atmosphere.[8] The extra carbon dioxide in the atmosphere causes additional warming, which in turn leads to more evaporation from the oceans. The added water vapor further amplifies the warming, because water vapor is also a greenhouse gas. To summarize, a small warming caused by the Milankovitch cycle initiates a chain of reinforcing feedbacks that lead to a much larger warming.

An opposite set of changes amplifies the cooling side of the Milankovitch cycles. When a cycle initiates a slight cooling, the cooling causes the oceans to absorb carbon dioxide from the atmosphere. This weakens the greenhouse effect, further cooling our planet and reducing evaporation from the oceans. The reduced evaporation means less water vapor in the atmosphere, amplifying the cooling until Earth plunges into an ice age.

Q I heard that the temperature changes in the ice core record precede the carbon dioxide changes. Doesn't this mean that you have cause and effect backward?

It is true that in some cases, temperature changes measured in the ice core record appear to have preceded a rise in carbon dioxide, but this does not

---

8 This release occurs because warmer water generally holds less dissolved gas, an effect you can confirm by popping open a warm can of soda, which will release gas more quickly than a cold can. Additional note: This fact may make you wonder why the oceans are currently absorbing carbon dioxide as our planet warms, and the answer is that it's a short-term gain arising from the great rate at which human activity is releasing carbon dioxide into the atmosphere. Ultimately, we expect the oceans to release more carbon dioxide as Earth warms, which will tend to exacerbate the problem of global warming over the long term.

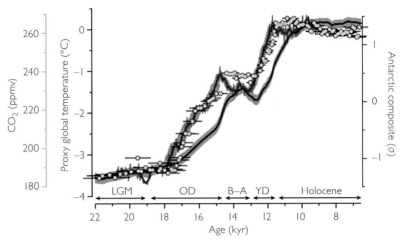

Figure 2.9 This graph shows a close examination of carbon dioxide concentration (yellow dots), Antarctic temperature (red curve), and global average temperature (blue curve) during the warming period that began about 20,000 years ago ("kyr" means "thousands of years"). Notice that while the Antarctic temperature changes come slightly ahead of the carbon dioxide changes, the global temperature changes do not. Source: J. D. Shakun et al., *Nature* 484 (Apr. 5, 2012): 49-54.

in any way change our understanding of the cause and effect. In fact, it's completely consistent with the idea that, as discussed above, the Milankovitch cycles trigger small changes in climate that are then amplified by feedbacks with carbon dioxide and water vapor. In other words, the feedback processes mean that once the Milankovitch cycles initiate a temperature change, both the temperature and the carbon dioxide concentration will rise or fall together, and at any given moment or any given place on Earth, one or the other may change first.

Further evidence that the cause and effect are well understood comes from a closer look at data from the end of the last ice age. The details are beyond our scope in this book, but the brief summary is as follows. The ice cores that show a slight lead in temperature changes compared to the carbon dioxide changes come from Antarctica, which means they reflect the temperature changes that occurred over the Antarctic ice sheet. However, scientists have other ways to study past temperatures, such as by drilling into sediments in lakes or the ocean floor, and these make it possible to measure past temperature changes in many places around the world. This work is fairly difficult compared to ice core measurement, but figure 2.9 shows what scientists found for the end of the last ice age. The yellow dots show the carbon dioxide concentration, the red curve shows Antarctic temperatures, and the blue curve shows global average temperatures from other measurements. Notice that while the Antarctic temperature rise came very slightly ahead of the carbon dioxide rise (which, as stated above, is unsurprising), the global temperature rise came *after* the carbon dioxide rise — completely undercutting any claim that cause and effect are backward.[9]

---

9 The delay between the carbon dioxide rise and the global temperature rise is thought to be due to complexities of ocean circulation; for details, see www.skepticalscience.com/skakun-co2-temp-lag.html.

What about the Medieval Warm Period, when Greenland was "green"? Doesn't that mean that we've been through much warmer periods in the recent past?

There are two questions here, so let's start with the first: It's true that the Vikings built settlements on the coast of Greenland beginning about a thousand years ago, during what is known as the *Medieval Warm Period* (roughly 950 to 1250 AD), when reduced Arctic sea ice made journeys to Greenland much easier than they were in the centuries before and after. But even at that time, Greenland was hardly "green"[10] — the vast bulk of Greenland has been covered by an ice sheet for at least several hundred thousand years. The Viking colonization never occupied more than a few coastal regions.

Now we turn to the second and more important question, which is whether the Medieval Warm Period is relevant to current global warming. The answer is a strong and definitive "no." The reason is simple: Even though there *was* a Medieval Warm Period, the amount of warming at the time pales in comparison to the warming going on today. Figure 2.10 shows the data from numerous independent scientific studies (each in a different color), along with recent temperature data (red). Notice that while the different studies do not all agree perfectly for times further in the past, they do all agree that today's temperatures are significantly higher than those of the Medieval Warm Period.

In fact, the evidence is even stronger than that shown in figure 2.10, because that figure shows only Northern Hemisphere temperatures. This is important because careful studies indicate that the Medieval Warm Period was a regional phenomenon that affected the northern Atlantic much more than other parts of the world; globally, there was little if any overall warming during this period. In other words, the Viking colonization of Greenland was made possible by regional, not global, climate changes. Today's warming, in contrast, is truly global.

 Wait — didn't I hear that the hockey stick graph has been discredited?

Well, you probably *have* heard this, since it is frequently repeated in places like the *Wall Street Journal*'s op-ed pages, *but it is not true*. The original

---

10   According to histories written not long after the colonization, the name "Greenland" was primarily a marketing ploy by the famed Viking Eric the Red, who believed it would encourage other people to make the journey there.

Northern Hemisphere Temperature Reconstructions

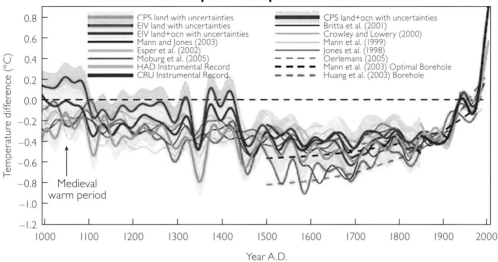

Figure 2.10 This graph shows more than a dozen different independent data sets all pointing to the same basic fact: Temperatures in recent years (solid red) have become significantly higher than they were during the Medieval Warm Period or any other time in the past 1,000 years. This graph is nicknamed the "hockey stick" because it looks kind of like a hockey stick lying on the ground with its tip pointing up at the right. Source: M. E. Mann et al., *PNAS* 105, no. 36 (2008): 13252–13257.

version of the "hockey stick" was published by climate scientist Michael Mann in 1998, and he used only a single data set. Skeptics jumped on it, claiming all kinds of reasons why the data should be doubted. Scientists took the skeptic concerns seriously, and therefore did what scientists do: They investigated in more detail. Indeed, the reason you see so many data sets — from independent sources including tree rings, corals, stalagmites, ice cores, and more — in figure 2.10 is that the scientific community went to great lengths in trying to either confirm or refute Mann's original "hockey stick." Keep in mind that every curve you see in figure 2.10 represents many years of fieldwork and careful research by a substantial group of scientists, who often put their lives on the line to collect the data in remote and dangerous locations. As you can see, these additional studies clearly confirm Mann's original conclusions. Still not mollified, the skeptics were so adamant in their objections that they convinced Congress to ask the National Research Council (NRC) to investigate those conclusions. The NRC report, published in 2006, concluded that the graph and the data were fully valid. Additional research since that time has only further strengthened the case for the validity of the "hockey stick" and what it tells us about changes in global temperatures over the past 1,000 years. If you want to learn more about this issue, two great sources are the NRC report (which you can download free at www.nap.edu/catalog/11676/surface-temperature-reconstructions-for-the-last-2000-years) and Michael Mann's book *The Hockey Stick and the Climate Wars* (Columbia University Press, 2012).

You've looked back 800,000 years, but I've heard that if you go back further, both carbon dioxide levels and temperatures were significantly higher than they are today. What do you say to that?

Although it's more difficult to figure out what the climate was like for times further in the past, evidence does indicate that there have been times when Earth was much warmer (and times when it was much colder) than anything we see in the 800,000-year record. For example, during much of the dinosaur period, evidence indicates that the global average temperature was significantly warmer than it is today, and the carbon dioxide concentration may have been above 1,000 parts per million, or more than double today's just-passed-400 parts per million. But I do not find these facts the least bit comforting; quite the opposite, as they seem to me to suggest that current warming could be even more devastating than most scientists generally assume.

Let's start by considering the implications of the warm temperatures of the age of the dinosaurs. These temperatures were high enough that there were no ice caps in either the Arctic or the Antarctic (figure 2.11), suggesting that a warming that brings back the temperatures of those times would cause the ice caps to melt completely. If that happened (a possibility we'll discuss in more detail later), sea level would rise so much that every coastal city in the world — not to mention most of Florida, Texas, and other low-lying coastal regions — would be deep under water. (Indeed, sea level during portions of the dinosaur period was more than 200 feet higher than it is today.)

As to the carbon dioxide concentration of the distant past, I'll make two points. First, you may hear skeptics claim that the fact that life thrived when the carbon dioxide concentration was much higher than it is today is proof that life can thrive under such conditions. Well, it is — but it's only proof for the species that were living at the time and therefore were adapted to those conditions. There is no reason at all to think that today's plants and animals would thrive similarly, because today's life is adapted to today's much lower carbon dioxide levels. It's far more likely that such a high concentration of carbon dioxide would cause severe damage to today's ecosystems. Second, while a carbon dioxide concentration of 1,000 parts per million sounds very high compared to today's 400, a look back at figure 1.10 shows that at the current rate of increase, we would surpass that level in just a few hundred years. Returning again to the quote by Margaret Thatcher, what we are doing to our planet today is unprecedented and "new in the experience of the Earth."

Figure 2.11 This painting shows Antarctica as it may have looked about 70 million years ago, when our planet was so warm that there were no polar caps and the carbon dioxide concentration was probably above 1,000 ppm. Source: Artwork by James McKay, University of Leeds, from V. Bowman et al., *Palaeogeogr., Palaeoclimatol., Palaeoecol.* 408 (Aug. 15, 2014): 26–47.

Skeptic Claim 3, Part 2: The Reliability of Models

A closely related claim holds that fears of global warming are overblown, because current models overestimate how the climate responds (the "climate sensitivity") to changes in the carbon dioxide concentration and therefore also overestimate the future warming. For example, most models predict that under a "business as usual" scenario (meaning no significant reductions in our current carbon dioxide emissions), the temperature would rise by 4–5°C (7–9°F) by the end of this century (figure 2.12), but some skeptics have claimed the rise won't be more than 2°C even in the worst case. In essence, they claim that the models overestimate climate sensitivity because they are missing key factors that might mitigate the future warming.

Could these skeptics be correct? The first thing that any scientist will tell you about modeling is that it's not easy. As an old saying goes, "Prediction is hard, especially about the future." But hard is not the same as impossible, and as we've discussed, today's sophisticated climate models do a good job of "predicting" the climate that exists today (meaning that the models match reality quite well, as you saw in figure 2.7).

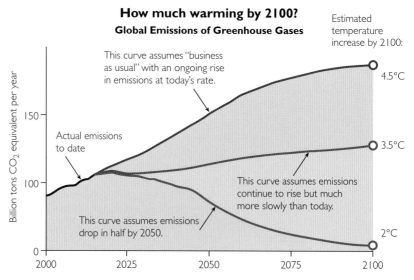

Figure 2.12 These three curves show three different possible scenarios for the rise or fall in human-caused emissions of carbon dioxide through 2100, and the numbers at the right show the temperature increase that models predict for each case. Notice that the "business as usual" curve leads to a global average warming of 4.5°C (8°F) by the end of this century. The middle curve shows what happens if emissions flatten out at about their current levels. Perhaps the most sobering is the blue curve, which shows that even if we reduce emissions to half of today's amount by 2050 and much less by 2100, we still end up with a temperature rise of 2°C, or about 4°F. Source: www.climateinteractive.org/tools/scoreboard/.

We can therefore have some confidence that these models should be at least modestly reliable in predicting what will happen in the future.

Notice my choice of words: "modestly reliable." The skeptics like to jump all over this type of honest assessment of the validity of models and essentially claim that we therefore shouldn't rely on the models at all, particularly when they predict dire impacts from global warming. But that's like saying that hard is impossible — it's just not true. It is possible that the models will prove to be far off the mark, but we certainly wouldn't expect that, based on how well they work for the present climate. So let's look in a little more detail at the common skeptic claims about modeling and see why the vast majority of scientists find these claims unconvincing.

Today's models can't even predict the weather more than a few days out, so how could they possibly predict the effects of global warming many years from now?

This question comes up frequently in the media, but it's based on a misconception about the nature of *weather* and *climate*. The two terms have very different meanings:

- *Weather* refers to the ever-varying combination of winds, clouds, temperature, and pressure that makes some days hotter or cooler, clearer or cloudier, or calmer or stormier than others.
- *Climate* refers to the *average* of weather over many years. For example, we say that a desert has a dry climate, even though it may sometimes rain or snow.

It is always easier to make predictions about long-term averages than about short-term variations. For example, when 10,000 people go to a casino, we can't predict the wins and losses for any single individual, but we know that on average, the gamblers are going to lose more than they win, which is how casinos make money. Similarly, we can't say whether or when a particular smoker will get lung cancer, but we know that on average, lung cancer is much more common among smokers. The situation with weather and climate is exactly the same: It's much more difficult to predict the short-term variability of weather than the long-term average that represents climate.

In other words, the fact that it's hard to predict the weather is *completely meaningless* in considering whether we can predict the climate. The evidence that we can predict the climate comes from the fact that today's sophisticated climate models do an excellent job of matching up with the present climate reality, giving us confidence that they should be similarly good at predicting the future climate.

It's also worth noting that when I say that today's models do an excellent job of matching reality, I mean much more than just the global average temperature. Today's models make regional predictions, and these regional predictions also match up to reality quite well. For example, as we'll discuss more in the next chapter, models have predicted numerous regional changes that appear to be occurring as expected, such as increases in drought in California, in storms along the East Coast, in flooding in regions of Asia, and much more. To ignore the insight provided to us by climate models simply does not make any sense.

OK, but maybe the models are missing an important mitigating factor. For example, aren't there uncertainties about the effects of clouds?

It's true that clouds are very complex and their effects are still not fully understood. This fact opens the way for a skeptic argument that basically goes like this: Global warming means the world starts getting warmer, and this increases evaporation from the oceans, which in turn means more clouds, which in turn reflect more sunlight, thereby stopping further warming. In other words, the claim is that clouds represent a negative feedback that acts to prevent global warming from getting too bad.

The problem with this argument is that it ignores two other effects. First, the additional evaporation that leads to the additional clouds also means there is more water vapor in the atmosphere, and recall that water vapor tends to amplify the effects of changes in the carbon dioxide concentration. While there is some legitimate debate over which effect — the extra heating from water vapor or the extra cooling from clouds — is stronger on relatively short time scales, there is little doubt about water vapor's amplifying role over periods of decades or longer. Moreover, even for short time scales (years), current understanding of cloud physics has led most scientists to conclude that the heating effects are stronger than the cooling ones.

Second, clouds are not the only thing that matters here. What really counts is the total reflectivity of Earth: If the reflectivity increases, that tends to cool the planet, and if it decreases, it tends to warm it. And while clouds contribute to the total reflectivity, the surface also plays an important role, and global warming is causing the surface reflectivity to *decrease*, which tends to make the warming worse. The major reason for this decrease is the melting of ice, which is much more reflective than the water or ground cover that replaces it after it melts. For example, the fact that much of the Arctic ocean is now ice-free for much longer times each year than in the past means that this ocean reflects less sunlight (and therefore absorbs more), which leads to more heating.

Most climate scientists suspect that, together, the warming effects of added water vapor and reduced ice cover should overwhelm the cooling effects of additional clouds. So while you may still hear this argument raised on occasion, you should recognize that any claims that clouds will mitigate our problems are on very weak ground.

Still, given the complexity of the climate and the uncertainties of the models, isn't it reasonable to think that other feedbacks may mitigate the threat?

It's certainly conceivable that the models could still be missing other factors that might mitigate future warming (and also possible that they are missing factors that might amplify it). However, if any such mitigating factors exist, they would have to have some rather strange properties. In particular, because they would by definition be factors that are not considered by current models, they would have to be both *unimportant* enough that they haven't caused major failures in the model results through the present time and *important* enough to make a major impact on the model results for the future. We cannot completely rule out the possibility that such factors exist, but these odd properties make it seem rather unlikely. This is a major reason why the vast majority of scientists reject the skeptic claims and instead conclude that the threat of global warming is every bit as bad as the models suggest it to be.

What if I still don't trust the models?

OK, let's say you want to ignore all the evidence from the models that I cited above and stick to the skeptic claim that, because the models can never be perfect, we shouldn't trust them at all. Well, that's not very scientific of you, but let's go with it for the moment . . .

Whether or not you believe the models, you still have to make decisions about what, if anything, we should do about global warming for the future. If you don't want to consider the models, then the only viable alternative is to base your decisions on the actual data. So let's see what the data tell us. Look back at figure 2.8, where you can see how temperature and carbon dioxide have changed together in the past. Notice that:

- Over the past 800,000 years, the largest upward swings in carbon dioxide concentration have been from about 180 to 290 parts per million, which is an increase of about 60%.
- These 60% increases in carbon dioxide concentration have been accompanied by temperature increases of about 8°C to 10°C (14°F to 18°F).

Putting these two facts together, the past data seem to suggest that a 60% increase in the carbon dioxide concentration will cause warming of 8°C or more. This is substantially greater warming than any of the models suggest for the rest of this century, even though current trends mean we will have reached a doubling (100% increase) of the carbon dioxide concentration well before the century ends. In other words, the projection you would make from actual past data is *worse* than what the models are suggesting.

What's the bottom line on the possibility that temperatures won't rise as much as the basic science might make us expect?

As we've seen, if you are looking for a mitigating factor that might "save" us from the otherwise scary predictions about global warming, no such factor has been found, and the success of models to date makes it unlikely that any such factor exists. We cannot be certain that the model predictions are accurate, but we ignore them at our peril.

Skeptic Claim 3, Part 3: Benefits May Outweigh the Risks

Because there seems little way to deny the reality of global warming, some skeptics instead try to claim that this warming will be *good for us* and therefore that we do not need to do anything about it. This is a rather remarkable assertion, because these skeptics are essentially advocating that we continue doing an "experiment" on our planet without being sure of the consequences. It is especially surprising when you realize that many of the people taking this position claim to be great admirers of conservatives like Ronald Reagan and Margaret Thatcher, yet they are clearly violating Reagan's dictate about the common sense of preserving (not dramatically changing) the environment (see quote that opens the introduction) and ignoring Thatcher's warning that we are "changing the environment of our planet in damaging and dangerous ways." Nevertheless, let's take a brief look at a few of these recent claims that global warming will be beneficial.

Could the added carbon dioxide help plant growth and agriculture?

One of the key "benefits" claimed for global warming is that the increased carbon dioxide concentration will increase plant growth, thereby helping agriculture. As I noted earlier, the skeptics making this claim often point to the thriving plants of dinosaur times, when carbon dioxide concentrations were far higher than they are today. But we've already seen how this argument falls apart. Plant and animal species adapt over time to the environments in which they live. The species that thrived in dinosaur times had millions of years to adapt to those conditions, while today's plants and animals are adapted to today's conditions.

Skeptics also point to small-scale experiments that have shown limited benefits for crops such as soybeans and rice with higher carbon dioxide concentrations. But they ignore the overall ecological effects that may be far more important. Remember that plants and animals are adapted to *local* climates. If the climate changes slowly, then species can adapt or migrate to survive. But if the climate changes faster than they can adapt or migrate, then they will die out or be replaced by other species. The rapid rate of climate change today therefore makes it likely that there will be great changes in the distribution of the plant and animal species living around the world, which means great changes to the entire ecosystem upon which our modern economy depends. While there may be a small chance that all these changes could actually prove beneficial, the vast majority of scientists suspect that the changes will be detrimental. You'd have to be an audacious gambler to be willing to continue on the path we are on in the small hope that it will turn out to be beneficial in the end.

Won't the melting of Arctic ice be good for us in opening up the Arctic sea?

The melting of Arctic sea ice is already causing a rush for Arctic riches, such as faster shipping routes among northern countries and access to minerals, oil, and other resources of the Arctic Ocean. By themselves, these things would seem to be beneficial to the global economy. But the key words are "by themselves," because they don't occur in isolation. Instead, they are consequences of having less sea ice, and that appears to be a very detrimental development on at least two levels.

First, as I've already noted, the fact that water is less reflective than ice means that Earth absorbs more heat from the Sun when Arctic ice melts, and this will only exacerbate the effects of global warming. Second, the distribution of ice in the Arctic has very significant effects on regional and global weather patterns, so we can expect major changes in atmospheric circulation and local weather as the Arctic melts. Already some scientists suspect that, through a complex set of interactions, the reduced amount of *summer* ice in the Arctic may be linked to the extreme *winter* weather that has affected the United States and Europe in recent years. While there's great debate among scientists about whether this particular linkage is real, we should expect at least some significant weather consequences from changes in the amount of sea ice.

The bottom line is that the melting of Arctic ice essentially is yet another unprecedented "experiment" on our planet. While there's always a small chance that this experiment will result in more pros than cons, it seems a highly dangerous experiment to conduct.

So how much credence should I give to the skeptic arguments overall?

A few decades ago, some skeptics tried to claim global warming wasn't real at all. As the evidence accumulated, they tried to claim it was a natural change rather than human caused. Now, recognizing that they can no longer legitimately dispute the human causation, they've turned to claims that global warming might not prove to be so terrible, or might even be beneficial to us. In every case, they've laid out arguments that other scientists have shot holes through, and in careful consideration of all the evidence, the vast majority of scientists have rejected the skeptic claims. Indeed, as has been widely reported, surveys have found that more than 97% of scientists who have spent their lives studying the climate accept the "consensus" view and reject the skeptic claims.[11]

In response to this last fact, the skeptics like to point out that science is not a democracy, and scientific facts can't be changed by a vote. But if you're not a scientist, you still have to decide who to trust. Imagine that

---

11   Multiple surveys have found this approximately 97% value, as summarized at www.skepticalscience.com/global-warming-scientific-consensus-advanced.htm and published in J. Cook et al., *Environ. Res. Lett.* 11, no. 4 (April 13, 2016).

you visited 100 doctors, and 97 of them told you, "You need to eat less sugar," while three said, "More sugar is good for you." Which would you believe? I'll let you answer that question for yourself, but note that you should give exactly the same answer to the question of who you should believe about the potential dangers of global warming.

Skeptic Claim 4: It's Warming Up, Humans Are Causing It, It's Harmful, But It's Not Cost-Effective to Solve It

We've now addressed all the major skeptic claims that relate directly to the science. However, another group of skeptics takes a different tack. This group accepts that global warming is real, human caused, and serious — but argues that the costs of dealing with it are high enough that more good could be done by applying our efforts in other areas. The best-known advocate of this idea is the Danish writer Bjørn Lomborg, but similar ideas have been advanced by many others.

Because this claim is based more on economics than on science, it's less of a clear-cut call. For example, as those behind these claims frequently remind us, the low-cost energy of fossil fuels has been a key to strengthening our modern economy and raising millions of people out of poverty. If we are going to move away from fuel sources that have had so many clear benefits in the past, we ought to have a really good reason. I think we do . . . but in order to make that decision, we need a clear understanding of the risks posed by global warming and the costs of alleviating those risks. We'll turn to these issues in the next two chapters.

3 The Expected Consequences

We served Republican presidents, but we have a message that transcends political affiliation: the United States must move now on substantive steps to curb climate change, at home and internationally. There is no longer any credible scientific debate about the basic facts: our world continues to warm, with the last decade the hottest in modern records, and the deep ocean warming faster than the earth's atmosphere. Sea level is rising. Arctic Sea ice is melting years faster than projected.

—William D. Ruckelshaus, Lee M. Thomas, William K. Reilly, and Christine Todd Whitman, former heads of the Environmental Protection Agency under Presidents Nixon, Reagan, George H. W. Bush, and George W. Bush, Aug. 1, 2013 (statement in the _New York Times_)

If you still have any doubts about the fact that global warming is _not_ a partisan issue, the quote above should dispel them. It is from four Republican leaders of the Environmental Protection Agency who served in the past four Republican presidential administrations. You will not find any substantial difference between the urgency that these Republican administrators bring to the issue and what you would hear from recent Democratic administrators, or from President Obama or former Vice President Al Gore. Global warming transcends politics, and it transcends international borders.

To understand why the issue requires urgent action, we need to understand in more detail the types of consequences expected if we don't act swiftly to curb the emissions that are causing global warming. By itself, the predicted rise in global average temperature of 2°C to 5°C (4°F to 9°F) by the end of this century might not sound so bad; it might even sound pleasant if you live in a cold climate today. But the expected consequences go far beyond a change in average temperature.

What is the basic science behind the consequences of global warming?

You've probably heard great debate in the media about the consequences of global warming, such as whether it has caused recent weather events. But while there is room for legitimate debate over particular consequences, the basic science behind the consequences is easy to understand. We can see it with a simple chain of logic:

> ***Starting point:*** As we've discussed, the underlying cause of recent global warming is human-caused emissions of carbon dioxide and other greenhouse gases.

>> ***Recall that:*** About half of this carbon dioxide is staying in the atmosphere, where it is responsible for the increasing carbon dioxide concentration (see figure 1.10), while much of the rest is being absorbed by the oceans.

> ***In the atmosphere:*** The rising concentration of carbon dioxide and other greenhouse gases strengthens the greenhouse effect (see figure 1.4), and a stronger greenhouse effect means more total energy trapped in Earth's atmosphere.

>> ***Therefore:*** Although we usually focus on warming, we should expect the consequences of the stronger greenhouse effect to include anything that can result from added energy in the atmosphere and oceans. Besides overall warming, these consequences can include changes in regional climates, more powerful storms and other extreme weather events, and melting of ice both on land and in the oceans.

> ***In the oceans:*** The added carbon dioxide dissolves in the water, where chemical reactions make the water more acidic.

>> ***Therefore:*** Another effect of global warming is what we call *ocean acidification*, which can cause great damage to coral reefs and other ocean ecosystems.[1] These changes, combined with pollution and overfishing, are likely to disrupt the food chain critical to sustaining the global fish stocks on which billions of the world's people rely for food and livelihoods. The ecosystem changes in the ocean may also have feedback effects on other changes arising from global warming, potentially amplifying many other effects on human civilization.

Figure 3.1 summarizes this logical chain. In the rest of this chapter, I'll discuss the major consequences identified above.

---

1   As I briefly noted earlier, global warming also appears to cause an increase in low-oxygen regions in the oceans, and some scientists worry that this may prove even more damaging to ocean life than ocean acidification. However, I will not discuss this effect here, because it is less well understood.

Expected Consequences of Global Warming

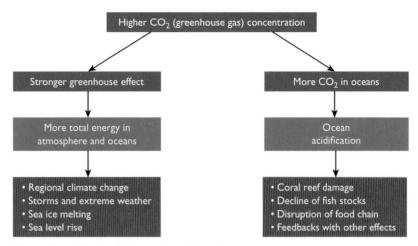

Figure 3.1 This simple logical chain (flow chart) explains why we expect a variety of consequences from what we usually just call *global warming*.

Q Can you give us an analogy to explain why the extra energy trapped by a stronger greenhouse effect will do more than just warm the planet?

Consider heating a pot of water on a stove. The energy from the stove obviously heats the water, but that's not all that it does. For example, some of the energy makes the gas flame or electric coil glow, some of it goes to the surrounding air, and some of it even drives "weather" (more technically, *convection*) in the pot by causing the water to circulate as heated water rises up from the bottom and cooler water sinks down from the top. In the same way, the added energy trapped by a stronger greenhouse effect in the atmosphere not only heats the air but also heats the oceans, melts ice, and drives more weather (and extreme weather).

Q How much energy is global warming adding to the atmosphere?

It's a surprisingly large amount. A recent analysis[2] found that at the current rate at which human activities are emitting carbon dioxide and other greenhouse gases, we are causing the total energy of the atmosphere and oceans to increase by approximately 250 trillion joules each second. To put this in more concrete terms, that is the equivalent of, for example:

- The energy that would be added by detonating four Hiroshima atomic bombs each second.

---

2 The numbers and most of the comparisons here are from 4hiroshimas.com, based on the research paper by D. Nuccitelli et al., *Phys. Lett. A* 376, no. 45 (Oct. 1, 2012): 3466–3468. You can also read about this work at skepticalscience.com/nuccitelli-et-al-2012.html.

- The energy that would be released by 500,000 lightning bolts each second.
- Enough energy to have two places in the world being struck by a hurricane as powerful as Hurricane Sandy at all times.
- Enough energy to power almost 3,000 tornadoes of the most destructive category (F5) each day.

With that much energy being added to the atmosphere and oceans,[3] it should not be at all surprising that we would expect noticeable consequences. Indeed, you might wonder why the consequences aren't even greater than they are, and the answer is that most of the energy is going into the gradual warming of the oceans and atmosphere.

Regional Climate Change

You've probably noticed that the terms "global warming" and "climate change" are often used interchangeably. There's a reason for this: A warming of Earth's average temperature is expected to mean regional changes in climate that can be much more or much less than average. They just have to all average out. In other words, while Earth as a whole may warm by "only" 2°C to 5°C during this century, different regions may experience more dramatic changes in climate, and these changes will also have numerous secondary impacts on our lives. This regional climate change is already under way, and we can expect greater change in the future.

Q   What evidence shows that regional climate change is already occurring?

The most direct way to look at regional climate change is in terms of temperatures. Figure 3.2 compares average regional temperatures over the five-year period from 2011 to 2015 to the averages for the period 1951–1980. Notice that almost all regions of the world have been significantly warmer in recent years than they were a few decades earlier, but some regions — particularly in the far north — have warmed much more than others.

There are many other ways to see the climate change that is under way, but I probably don't need to tell you that. By now, the changing climate is fairly obvious to most people, with frequent news reports of weather records being broken while reporters interview long-time

---

3   A few details of the comparisons: The first two are simply other ways of stating the equivalent amount of energy per second. The third is based on the fact that the energy increase (per second) is equivalent to the energy released (per second) by two major hurricanes. The fourth is based on taking the energy released each *day* and calculating how many tornadoes would be represented by the same total energy.

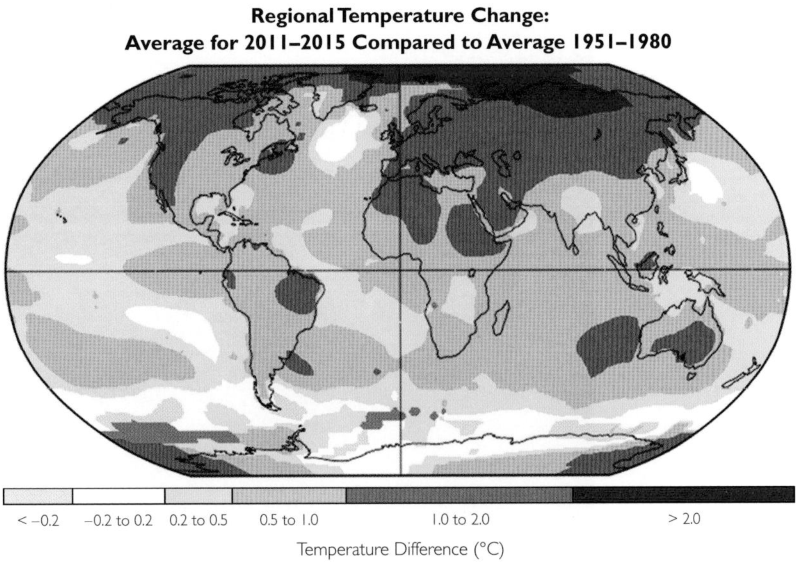

**Regional Temperature Change:
Average for 2011–2015 Compared to Average 1951–1980**

| < −0.2 | −0.2 to 0.2 | 0.2 to 0.5 | 0.5 to 1.0 | 1.0 to 2.0 | > 2.0 |

Temperature Difference (°C)

Figure 3.2 This map shows regional temperature changes, comparing the five-year period 2011–2015 to the average from 1951 to 1980. Notice that warming has affected almost all places on Earth, but some areas have warmed much more than others. (Gray areas represent a lack of data for the comparison.) NASA has produced an outstanding video of these changes, looking all the way back to 1880, which I've posted at https://youtu.be/4njRzbPdD34. Source: NASA, generated at data.giss.nasa.gov/gistemp/maps/.

residents saying they have never seen anything like the hurricane, blizzard, flood, or fire they've just experienced. Nevertheless, data like those shown in figure 3.2 prove it beyond any reasonable doubt: Climate change is already happening.

Besides the temperature changes, what else happens as a result of regional climate change?

There are too many secondary effects to list fully, but here are a few examples:

- Drought and floods: The changes in weather patterns cause some regions to become drier and others to become wetter. For example, the recent drought in California (California's Sierra Nevada snow-pack for 2015 was the lowest in at least 500 years) and the American Southwest is likely worse than it would have been without climate change, and the same is true for regions (such as in Pakistan) that have had increased flooding in recent years. You can understand these changes by recalling that warmer temperatures cause more evaporation of water. This extra evaporation tends to make dry

Figure 3.3 The Colorado High Park wildfire may or may not have been tied to regional climate change, but we expect fires of this type to become more common in regions that become drier and hotter over time. Source: U.S. National Forest Service.

places even drier and therefore more prone to drought, while at the same time making more total moisture available to fall as rain (or snow), thereby increasing the likelihood of floods.

- Wildfires: Drier and hotter conditions in turn lead to greater danger from wildfires (figure 3.3), and significant increases in wildfires already have been seen in many places around the world, including the American West, Alaska, Canada, Australia, and Russia.
- Dying forests: The pine forests of the Rocky Mountains are currently dying off due to an explosion in the population of pine beetles that kill the trees. This increase in the pine beetle population has been tied to the shorter and warmer winters that are a result of climate change (figure 3.4). Similar ecological effects are happening in many other regions around the world.
- Pests and diseases: Global warming is probably contributing to increases in the spread of insect-borne illnesses and crop-damaging pests. For example, diseases such as Zika, dengue fever, and chikungunya are transmitted by mosquitoes, which pick up these diseases by biting infected people and then, after some incubation period, spread them to other people through subsequent bites. Research has shown that warmer temperatures shorten the incubation period, which means the disease can spread more rapidly. Warmer

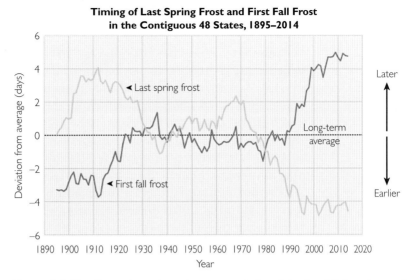

Figure 3.4 This figure shows how the average timing of the first fall frost (orange) and last spring frost (yellow) has changed over the decades compared to the long-term average. Notice that the first fall frost is now coming more than four days later and the last spring frost more than four days earlier, for a net of eight additional frost-free days compared to the past average. The shorter frost season has been tied to the spread of pine beetles and other pests and diseases. Source: U.S. Environmental Protection Agency, based on a 2015 update to data from K. E. Kunkel et al., *Geophys. Res. Lett.* 31:L03201.

temperatures also tend to increase the geographic ranges of insects, which is thought to be a major reason why insect-borne illnesses are spreading into regions that were once free of these diseases.

- Species extinctions: Some species cannot migrate fast enough as local climates change, causing them to go extinct as the ecosystems they rely upon are altered or disappear.

What future effects can we expect from regional climate change?

In brief, we expect that as the total amount of global warming continues to increase, future regional effects of climate change will essentially be magnified versions of what we are seeing already. The amount of magnification depends on how much the globe warms overall. For example, if you live in a region that has experienced increased drought or increased wildfires in recent years, you can expect these events to become even more common during the rest of this century. Similarly, if you are in a region that has experienced more flooding in recent years, massive floods that used to happen once in a generation may become your "new normal."

3

The Expected Consequences

I've heard concern about thawing of permafrost; could that amplify these effects?

As if the consequences described above (and below) aren't bad enough, some scientists think there may be cause for even greater alarm: Vast amounts of both carbon dioxide and methane are currently stored in Arctic regions in *permafrost*, by which we mean ground or tundra in which temperatures generally remain below freezing all year round. The permafrost contains the remains of plants that have not decayed because of the freezing temperatures. If warming causes this permafrost to thaw, we might expect the material to decay and release its carbon dioxide and methane into the atmosphere. In that case, the concentrations of both of these greenhouse gases might rise even more dramatically and rapidly than they are already rising, effectively amplifying all the effects of global warming. I won't say much more about this, but it's worth keeping in mind that because of amplifying possibilities like this one, what we often think of as "worst-case" scenarios might not truly be the worst cases.

Storms and Extreme Weather

A second major consequence we can expect from global warming is more extreme weather events. As we've discussed, global warming really means an increase in energy in the atmosphere and oceans, and energy is what drives weather. With more energy, we expect hurricanes, thunderstorms, and other extreme weather events to become more numerous, more severe, or both. Notice that extreme events include severe winter weather, leading to the ironic result that global warming can lead to heavy winter snowfalls.

 Can we tie particular storms to global warming?

No, we cannot tie any particular storm to global warming. However, we can tie an overall trend to global warming. Many climate scientists use loaded dice as an analogy. Just as loading dice makes certain outcomes more likely than they would be by natural chance, global warming makes extreme weather events more likely than they would be otherwise. Consider 2015's Hurricane Patricia (figure 3.5) as an example. This storm was one of the strongest hurricanes ever recorded, and it occurred during the second most active overall Pacific hurricane season on record. We cannot say that global warming caused this storm or this season. What we *can* say is that global warming makes storms like this and seasons like this more likely, and that we can therefore expect

Figure 3.5 Hurricane Patricia in October 2015 was one of the strongest hurricanes ever recorded and came during the second most active Pacific hurricane season on record. This photo was taken by astronaut Scott Kelly from the International Space Station. Source: Scott Kelly, NASA.

more years like this in the future. Cigarette smoking offers another analogy: We can't be sure that smoking caused a particular person's lung cancer, but we know that on average, the more you smoke, the more likely you are to get lung cancer. In the same way, as we add more greenhouse gases — and hence more energy — to the atmosphere, we should expect more extreme weather events.

Are we certain that extreme weather events are increasing?

It's not easy to define "extreme," and changes in human conditions can make storms that would once have been benign (for example, because they affected unpopulated areas) seem more severe simply because there are now more people in their path. This naturally introduces some uncertainty into statistics concerning extreme weather events. Nevertheless, the data strongly indicate an upward trend. Figure 3.6 shows a graph of recent natural catastrophes compiled by a leading insurance company. Aside from the red portions at the bottom, which represent nonclimate events such as earthquakes and volcanoes, the bars represent events that are linked to weather and climate. Notice that while there is quite a bit of variability from year to year, there's been a marked overall increase from 1980 to the present, providing strong evidence for the "loaded dice" analogy for how global warming makes extreme weather more common.

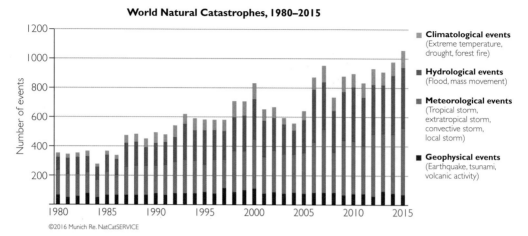

Figure 3.6 This graph shows the change in world natural catastrophes from 1980 through 2015. You can ignore the red portions of the bars at the bottom, which represent geological events (earthquakes, volcanoes) that are not affected by global warming. But all the other bar colors represent weather- or climate-related events. Although there is significant variability from one year to the next, the overall trend is clearly upward. Source: Munich Reinsurance Company, Topics Geo (2016 Issue).

We've had some extremely cold winters lately; doesn't that undermine the argument for global warming?

Not at all! Remember that storms are driven by energy in the atmosphere and oceans, and global warming means more energy. So storms of all types — *including winter storms* — can become more severe. Indeed, figure 3.7 shows evidence of a trend toward "when it rains, it pours" (or "when it snows, it blizzards") over recent decades, which is again consistent with what we might expect with global warming.

Should we expect heat waves to become more common than cold spells?

Yes, and the evidence shows this to be the case, both nationally and globally. Figure 3.8 shows how the ratio of record high temperatures to record low temperatures has changed by decade since the 1950s. Notice that while record lows outpaced record highs in the 1960s and 1970s, in recent decades the record highs have significantly outpaced the record lows — a clear indication that hotter weather is becoming more common. Keep in mind that this is not just uncomfortable but dangerous: While they tend not to make as much news as major storms, heat waves kill more people globally than any other type of weather event.

Melting of Sea Ice

It's fairly obvious that heat causes ice to melt, and therefore that we should expect global warming to contribute to a reduction in ice cover

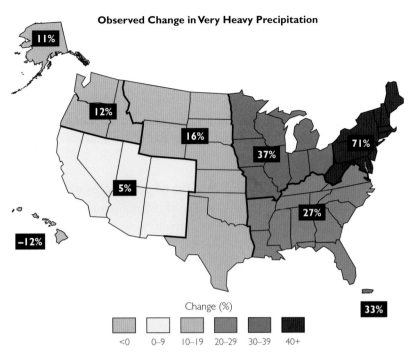

Observed Change in Very Heavy Precipitation

Change (%)

| <0 | 0–9 | 10–19 | 20–29 | 30–39 | 40+ |

Figure 3.7 This graph shows the percentage change in heavy rain and snow events for different regions of the United States from 1958 to 2012. Notice that in all regions (except Hawaii), the trend has been toward heavier events; in other words, when it rains, it pours (and when it snows, it blizzards). Source: U.S. National Climate Assessment 2014 (nca2014.globalchange.gov).

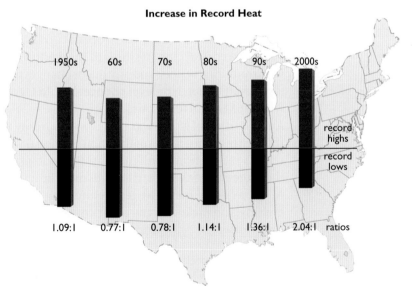

Increase in Record Heat

| 1950s | 60s | 70s | 80s | 90s | 2000s |

record highs

record lows

| 1.09:1 | 0.77:1 | 0.78:1 | 1.14:1 | 1.36:1 | 2.04:1 | ratios |

Figure 3.8 Notice the increase in recent decades in the ratio of record high to record low temperatures for the 48 contiguous United States. This is a clear indication of a trend toward more hot-weather extremes. Similar data support the same conclusion globally. Source: National Center for Atmospheric Research, based on data from G. A. Meehl et al., *U.S. Geophys. Res. Lett.* 36:L23701 (2009).

3

The Expected Consequences

around the world. Broadly speaking, there are two categories of ice melt, each with a different set of consequences:

1. The melting of sea ice, like that of the Arctic Ocean.
2. The melting of *glacial ice*, by which we mean landlocked ice such as that in Greenland and Antarctica.

We'll focus on the melting of sea ice as our third major consequence of global warming, saving the melting of glacial ice for the discussion of sea level that follows.

Why is the decline of sea ice likely to be detrimental?

The good news is that the melting of sea ice does not affect sea level, because this ice was already floating. The bad news is that it has many other detrimental effects. Most famously, it hurts polar bears, which depend on the ice in order to hunt seals, but it also affects weather patterns and has been implicated in changes in the jet stream and the so-called polar vortex that may have caused some of the strange weather in the U.S. in recent years. But these aren't even the most serious consequences.

The two greatest threats posed by the melting of sea ice are these:

1. Changes in ocean salinity (the amount of salt in the water): Melting ice adds fresh water to the oceans[4] and therefore lowers the salinity of the ocean water in the regions where the ice melts. The lower salinity may in turn lead to changes in ocean currents and the productivity of fisheries. No one knows exactly how damaging these changes may prove to be, but at the extreme, they could be very dangerous. For example, changes in ocean currents could have dramatic effects on coastal climates around the world and on nutrient levels in surrounding waters, and any major changes in fishery production (as a result of nutrient level changes) could leave billions of people without a critical food source.
2. Amplification of global warming: As I've already noted (see page 48), the melting of sea ice actually amplifies global warming, because ice reflects much more sunlight than water. Replacing ice with water therefore means that Earth absorbs more heat from the Sun. In other words, as sea ice melts, we get a reinforcing feedback that accelerates melting and makes all the other consequences of global warming even worse.

Why doesn't melting sea ice affect sea level?

Because the weight of floating ice already contributes to sea level, and its weight does not change as it melts. You can easily prove this for yourself by

---

4 Sea ice is generally fresh water (or at least much fresher than normal seawater) as a result of the way in which it freezes, which tends to push its salt out during the freezing process.

Declining Arctic Sea Ice (Septembers)

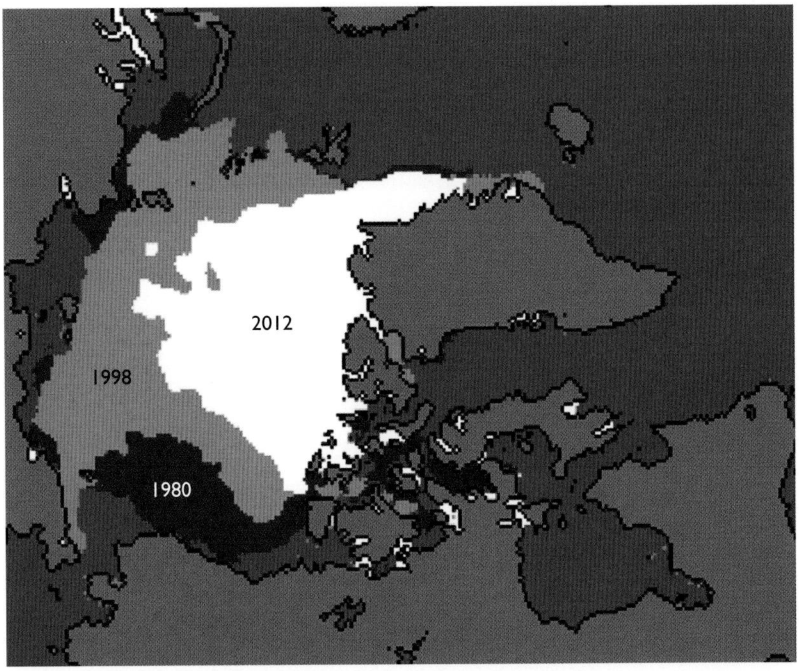

Figure 3.9 This map shows the declining extent of Arctic sea ice in September of three selected years. In 1980, the sea ice extended all the way to the edges of the red boundaries, including all the pink and white areas. In 1998, it extended only as far as the pink boundaries, and in 2012 only over the white region. Notice the huge decrease in the ice-covered area. Source: National Climate Assessment 2014, based on data from the National Snow and Ice Data Center.

grabbing a cup of ice water (in which the ice is all floating). Mark the water level when you start and when the ice has all melted, and you will see that it does not change.

What evidence indicates that sea ice is declining in the Arctic Ocean?

Figure 3.9 shows the September ice coverage in the Arctic Ocean for 1980, 1998, and 2012. Note the clear and dramatic decline. The September change is shown because that is the month when sea ice is generally at its minimum after the summer melting, but similar results are found for other months. (We must compare the same month in each year, since obviously the ice coverage grows in winter and declines in summer.)

Q You've shown the sea ice for 2012, which was a record low year. Is that fair?

Any time someone only shows selected data, as in figure 3.9, there's always a risk that he or she has "cherry picked" the data to show a particular result,

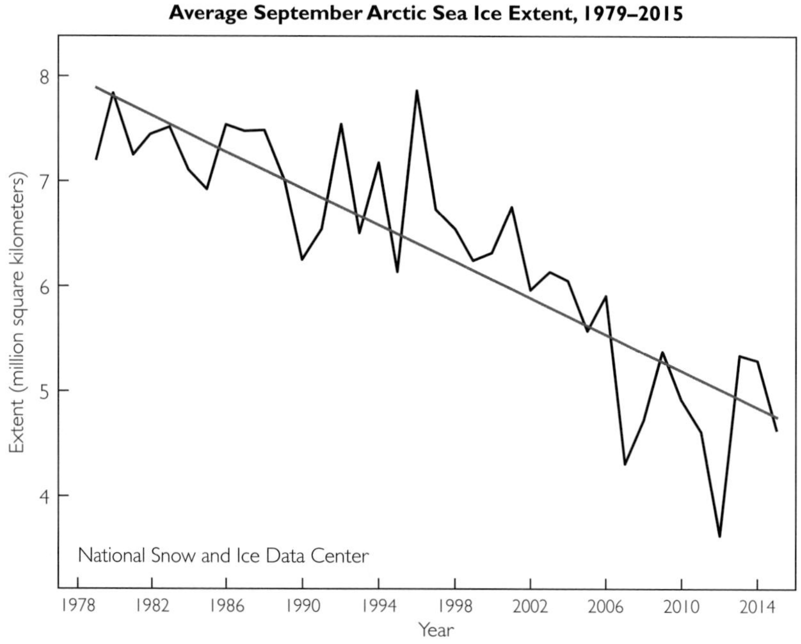

Average September Arctic Sea Ice Extent, 1979–2015

Figure 3.10 This graph shows the change in the total area of the Arctic covered by sea ice in September of each year (when the sea ice is near its minimum after summer melting) since satellite records have been available. The black curve shows the actual data, and the blue line is a "best fit" that shows the declining trend. The average rate of decline for the more than three-decade period has been more than 13% per decade, and the nine lowest September ice extents (over the satellite record) have all occurred in the last nine years through 2015. Source: National Snow and Ice Data Center. (Latest monthly data available at nsidc.org/arcticseaicenews/.)

while other data might show something different. The way to tell whether the overall data are being fairly represented is to look at the larger data set. Figure 3.10 shows the year-by-year data for Septembers from 1979 through 2015. You can see that the extent of sea ice was indeed lower in 2012 than in the next three years, but the overall downward trend is still very clear. Moreover, notice that the nine lowest years on record for ice coverage have been the last nine years, a result that is statistically consistent only with a significant downward trend.

Sea Level Rise

A fourth major consequence of global warming is an expected rise in sea level. This rise actually comes from two distinct processes. First, although we don't usually notice it, water expands very slightly as it gets warmer, and this *thermal expansion* has already been implicated

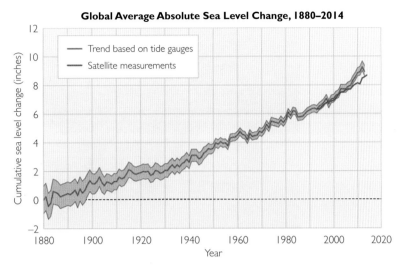

Global Average Absolute Sea Level Change, 1880–2014

Figure 3.11 legend:
— Trend based on tide gauges
— Satellite measurements

y-axis: Cumulative sea level change (inches)
x-axis: Year

Figure 3.11 This graph shows measurements of the overall rise in sea level since 1880, which is thought to be due primarily to thermal expansion of the water as ocean temperatures have risen. The shaded region shows the uncertainty range for the data. Note the overall rise of more than 8 inches, or 20 centimeters. The same effect is expected to increase sea level by up to another foot (30 centimeters) by 2100. Source: U.S. Environmental Protection Agency, based on data from NOAA and the Commonwealth Scientific and Industrial Research Organisation.

in a measurable rise in sea level. The second, and potentially greater, contribution to rising sea level comes from melting of glacial ice, particularly in Greenland and Antarctica, which adds water to the oceans.

How much is sea level rising due to thermal expansion?

Figure 3.11 shows measurements of the change in sea level since 1880, indicating an overall rise of more than 8 inches (20 centimeters). Based on scientific understanding of how sea water expands with rising temperatures, it is thought that much of this increase has been due to thermal expansion (as opposed to ice melting). Assuming that global warming continues as predicted, ongoing thermal expansion is expected to cause a rise of another foot, or 30 centimeters, by 2100.

A sea level rise of a foot may not sound like much, but it is enough to cause flooding in many low-lying regions around the world. Moreover, its effects can be magnified by storms, causing storm surges to rise higher and go farther inland than they would otherwise. Indeed, many scientists suspect that the tremendous damage from 2012's Hurricane Sandy (figure 3.12) was significantly magnified by the rise in sea level that has already occurred.

3
The Expected Consequences

Figure 3.12 This photograph shows damage in New Jersey from the Hurricane Sandy storm surge (2012). Source: Wikipedia/U.S. Air Force Master Sgt. Mark C. Olsen.

Q Why do the satellite data and tidal data differ in figure 3.11?

Toward the right side of figure 3.11, you can see that the satellite data show a smaller increase in sea level than the tidal data. This is not an error or any cause for concern, but a result of the fact that the "sea level" we observe along a coastline can be affected by at least two different processes: (1) an increase in the level of the ocean water, and (2) a decrease in the level of the land on the coastline relative to the ocean basin. Because tidal gauges measure sea level relative to the land level along coastlines, they account for both processes. In contrast, the satellite data measure only the actual height of the ocean surface, which explains the small difference in the data sets.

Q How much will sea level rise due to melting ice?

The second contribution to rising sea level comes from melting of glacial ice, particularly in Greenland and Antarctica. Scientists cannot yet predict ice melting very well, but many experts expect it to increase sea level by at least one meter (three feet) — and possibly much more — by the end of this century. This would have severe consequences along coastlines. For example, the red in figure 3.13 shows coastal regions of the southeastern United States that would be flooded by a one-meter rise in sea level, and the yellow shows regions that would be flooded by a six-meter rise.[5]

Also keep in mind that while figure 3.13 shows only the southeastern United States, similar effects are expected globally. Some island

5 The *New Yorker* magazine recently ran a great article about how sea level rise is already affecting Miami; you can find it at www.newyorker.com/magazine/2015/12/21/the-siege-of-miami.

Regions that will be below sea level

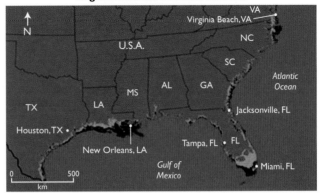

Red = 1-meter rise; yellow = 6-meter rise

Figure 3.13 While there are great uncertainties in estimating how rapidly sea level will rise, many experts expect the rise by 2100 to be at least about one meter (three feet), and a worst case might be as high as about 6 meters (20 feet). This map shows regions of the southeastern United States that would be flooded by a one-meter rise in sea level (red) and a six-meter rise in sea level (yellow). Other coastal regions around the world would experience similar inundation. Source: University of Arizona, Department of Geosciences, Environmental Studies Laboratory. If you want to see how sea level rise would affect other regions, there is a nice series of maps at www.globalwarmingart.com/wiki/Sea_Level_Rise_Maps_Gallery.

nations (and other inhabited islands) may end up completely underwater, and many of the nations that will be affected by sea level rise are much less equipped to deal with the consequences than a wealthy country like the United States. Many political scientists and military analysts fear that sea level change alone could displace hundreds of millions of people from their homes, leading to political upheaval on top of climate upheaval.[6]

Q I've recently heard people say that the Antarctic ice cap is actually growing, not shrinking. Are you sure that ice is actually melting into the sea?

What you heard is based on a study published in late 2015 (H. J. Zwally et al., *J. Glaciol.* 61, no. 230, pp. 1019–1036), and as I write in early 2016, scientists are actively debating the validity of the results. Nevertheless, even if the new study is correct, there's still no reason to doubt that ice is melting overall on Earth; the study only questions ice loss in Antarctica. Here's a brief summary of the issues.

---

6 As another example of security concerns raised by global warming, many analysts have linked the current turmoil in Syria and other regions of the Middle East to an extended drought in the region, which may have been exacerbated by climate change. Indeed, a 2014 report from the U.S. Department of Defense called global warming a "threat multiplier" that fuels terrorist groups.

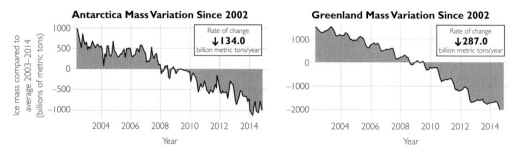

Figure 3.14 These data from NASA's GRACE satellites show changes in the mass of the Antarctic and Greenland ice sheets since 2002 (when the satellites were launched). Note that the rate of ice loss from Greenland is more than twice as large as the rate from Antarctica, which is presumably a result of the fact that temperatures have increased much more in the Arctic than in the Antarctic (see figure 3.2). Source: NASA (climate.nasa.gov/vital-signs/land-ice/).

There are two major methods used to measure changes in the Greenland and Antarctic ice sheets, both of which require satellite data. The first makes careful measurements of any changes in the strength of gravity over the ice sheets. These measurements should tell us how the total *mass* of the ice is changing with time (because the strength of gravity is determined by mass), and figure 3.14 shows the results from NASA's GRACE satellites. The second method looks for changes in the *elevation* of the ice sheets. The new study used this method and found that even though there was elevation loss in many parts of Antarctica (as expected if mass has been lost), there was enough elevation *gain* in other parts of the continent to suggest a net gain in the overall average elevation of the ice sheet.

The question, then, is how to reconcile the net gain in ice elevation claimed by the new study with the loss of ice mass measured by the GRACE satellites. One possibility, of course, is that one or the other of the measurement sets is wrong. For example, perhaps either the mass measurements or the elevation measurements have some calibration problem that makes them inaccurate. Most scientists have great confidence in the GRACE results, because they have been checked and rechecked for more than a decade now, but somewhat less confidence in the newer claims that have not yet been checked so carefully. There's also a way that both sets of measurements could be correct: Any elevation gain must be due to new snow that has fallen, which means it is at least possible that this snow is "fluffy" enough to have caused an overall elevation gain even while the total mass has declined. Climate scientists are working to determine which possibility is correct, which can make it difficult for nonscientists to know what to make of the debate. So for now, I'll suggest that you focus on three key points:

1. Even if the new study is correct, the amount of ice gain it claims for Antarctica is only about a third of the ice *loss* that GRACE has measured for Greenland. (Neither the new study nor any other studies have called the Greenland results into question.) In other words, any Antarctic ice gain is likely being more than made up for *globally* by Greenland's rapid ice loss.

2. The new study also found that the trend in Antarctica is toward smaller ice gains and predicts that it will result in net ice loss within a few decades at most. In other words, even if it is correct, this study at best

suggests that sea level rise will be less than otherwise estimated over the short term. That is, sea level rise will continue due to ice loss from Greenland and other sources, and it will accelerate once the Antarctic ice gain reverses.

3. This case is a great illustration of the way that science progresses through careful study in which results are checked and rechecked (and repeated or revised if necessary). This means it inevitably takes time for scientific debates to be resolved. It will take time for the details of this particular debate to be resolved, but this does not change the fact that 150 years of study have by now ensured that our basic 1-2-3 science is thoroughly understood and established. There is still much to learn about the precise consequences of global warming, but there is no denying that the basic problem is real.

What if the polar caps melted completely?

Up to this point, we've talked only about the expected sea level rise by 2100, but we can't expect the rise to stop there. If Earth warms enough, we might expect the polar caps to melt completely, returning our planet to the ice-free state of dinosaur times. So if you really want to understand the threat of sea level rise, we need to know how high it might go in this worst-case scenario. There are two general ways to answer this question. One is to look at geological data to see how much higher sea level has been in the past when Earth was ice-free. The other is simply to calculate the total volume of water in glacial ice and how much sea level would rise if that amount melted into the oceans. Both approaches yield the same dismaying answer: If the ice caps fully melt, sea level will rise by some 70 meters, or about 230 feet.

Although nearly all scientists suspect that such melting would take centuries or even millennia, it may already be inevitable[7] unless we find a way to *reverse* the effects of global warming (such as by removing carbon dioxide from the atmosphere so that its concentration drops back down). If we try to put the best face on this possibility, a time scale of centuries would in principle allow our descendants to migrate inland as the coastline shifts. Still, it suggests the disconcerting possibility that future generations will have to send deep-sea divers to explore the underwater ruins of many of today's major cities.

Should I sell my beach-front real estate?

I'm not in the business of giving investment advice, and if you sell before sea level rises too much, you might make out quite well. But one thing is near certain: If your plan is to keep your beach-front property in your family for generations, then your only hope of success lies in our rapidly stopping and then reversing the effects of global warming.

---

7 A pair of studies published in 2014 (one led by Eric Rignot, the other by Ian Joughin) concluded that the West Antarctic Ice Sheet is already doomed to long-term collapse, which by itself would raise sea level at least three meters. No one knows the "tipping point" for complete melting of the rest of the polar ice.

Ocean Acidification

The fifth and last major consequence we'll discuss is ocean acidification, which occurs as carbon dioxide dissolves in the oceans and, through a well-understood chain of chemical reactions, makes the oceans more acidic. As briefly noted in the introduction to this chapter, ocean acidification can have devastating consequences for ocean ecosystems, including killing coral reefs. It can therefore damage our civilization through both direct effects like reductions in fish stocks and indirect effects stemming from the way these changes in the oceans may affect other Earth systems, including the climate.

I won't say much more about ocean acidification, partly because it has been less well studied than other consequences and partly because the details require discussions of ocean chemistry and ecosystems that are beyond the level of this book. Nevertheless, it's important to keep in mind that because the oceans make up about three-fourths of Earth's surface, anything that affects the oceans is likely to affect the rest of the world as well. It is possible that ocean acidification could be as devastating in its effects on our civilization as our other four consequences combined, and it is therefore very important to consider it as part of any overall discussion of global warming.

What evidence indicates that the oceans are actually acidifying?

Direct measurements show the acidification (figure 3.15), leaving no doubt that it is real. The effects on coral reefs and other ocean ecosystems have also been observed and measured.

What's the bottom line for the expected consequences of global warming?

We have discussed five major categories of consequences of global warming: regional climate change, storms and extreme weather, melting of sea ice, sea level rise, and ocean acidification. We've found that each of these major consequences is likely to lead to a number of secondary consequences, and feedbacks among the various consequences could potentially make the whole of the problem worse than the sum of its parts. While there is uncertainty about exactly how serious all the consequences of global warming will prove to be, there is little scientific doubt that these consequences are real.

We've also seen that these are not just consequences for the future. Evidence shows that we are already feeling the effects of many of them, though we expect them to worsen in the future. The bottom line is that

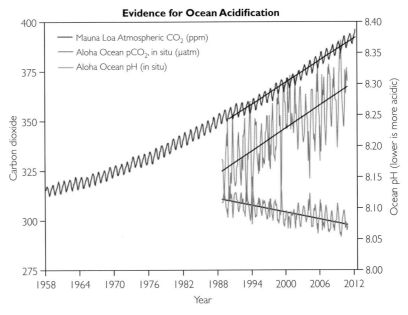

Figure 3.15 This graph shows evidence for ocean acidification. The red curve is a repeat of the atmospheric carbon dioxide rise from figure 1.8. The blue curve shows measurements of the carbon dioxide concentration in the ocean, which is rising along with the atmospheric concentration. The green curve shows the change in the ocean pH, which is a measure of the water's acidity; lower pH means greater acidity. The pH is going down as the carbon dioxide concentration goes up, demonstrating that the ocean is indeed becoming more acidic. Source: U.S. National Climate Assessment 2014 (nca2014.globalchange.gov).

the longer we continue adding carbon dioxide and other greenhouse gases to the atmosphere, the more detrimental to human life and civilization we should expect the consequences to be. With that in mind, it is time to turn our attention to what we can do about the problem before it gets completely out of hand.

4 The Solution

We have many advantages in the fight against global warming, but time is not one of them. Instead of idly debating the precise extent of global warming, or the precise timeline of global warming, we need to deal with the central facts of rising temperatures, rising waters, and all the endless troubles that global warming will bring. We stand warned by serious and credible scientists across the world that time is short and the dangers are great.

— Senator and Republican presidential candidate John McCain, May 12, 2008 (speech at Vestas Wind Energy Training Facility in Portland, Oregon)

I hope that by this point in the book, you'll understand and agree with the statement above by Senator John McCain. Moreover, while some people may debate Senator McCain's degree of conservatism, the fact that he won the Republican nomination for president should once again make clear that the science of global warming is *not* a partisan issue.

Of course, the fact that people of all political persuasions can agree on the nature of the problem does not necessarily mean that everyone will agree on the best way to solve it. A discussion of potential solutions therefore takes us away from the "pure science" focus of the previous chapters and into areas that are more a matter of opinion. For that reason, I'll admit to having had some reluctance about including this chapter in the book, because I can't defend everything in it with the same high level of evidence that I've presented in previous chapters. However, I also recognized that if I stopped at the end of the prior chapter, I would have been leaving you with lots of reasons to be concerned about the future without having given you any cause for great hope. And personally, I am very hopeful and optimistic about our future.

So with some trepidation about sharing my personal opinions in a book primarily focused on science, I'll go ahead and explain why I am so optimistic. I've already mentioned the basic reason in the introduction (see my third goal on page 2): I believe that if we go about this

Figure 4.1 This cartoon illustrates the idea that solutions to global warming are likely to be "win-win" in terms of their economic and political impacts as well as in alleviating the risks we face from global warming consequences. Joel Pett Editorial Cartoon used with the permission of Joel Pett and the Cartoonist Group. All rights reserved.

in the right way, there is a solution to all the problems of global warming that will not only protect the world for our children and grandchildren, but actually lead us to a stronger economy, with energy that is cheaper, safer, cleaner, and more abundant than the energy we use today. Moreover, I believe this solution can appeal to people across the political spectrum, and its "win-win" nature means it could even appeal to those few people who still don't believe we face a real threat (figure 4.1). After all, who can argue against something that alleviates the risks of global warming while also improving our economy and our lives?[1] In the rest of this chapter, I'll focus on technologies that can offer us replacements for the fossil fuels that cause global warming, on what I believe to be the major obstacle that has prevented a solution to date, and on what I consider the simple and obvious "win-win" solution that will transcend politics and benefit everyone, both in the United States and around the world.

---

1 Well . . . while no one is likely to argue against these things directly, any time our economy changes in a significant way — as it would in a transition away from fossil fuels — some people will be at risk of losing money even if the overall economy makes gains. In this case, those at risk are those with financial stakes in the current fossil fuel economy, so it's not a great surprise that some of these folks have spent enormous amounts of money trying to convince the public that global warming isn't real. Still, even in these cases, I think their concerns are misplaced: If they would take the same resources they use to argue against reality and instead direct them toward developing new energy technologies, they would be well positioned to succeed in a new energy economy.

Replacement Energy Technologies

The only sure way to alleviate the consequences of global warming is to stop adding carbon dioxide (and other greenhouse gases) to the atmosphere. Because most of our greenhouse emissions come from the burning of fossil fuels, this means we must find a way to replace these fuels with other energy sources. This brings us to good news: Technologies that would allow us to completely stop our use of fossil fuels already exist, and future technologies have even more promise. I'll start by discussing the existing technologies that I believe can, in some combination, offer a solution: energy efficiency, renewable energy from sources such as wind and solar, and nuclear power.

What is the role of energy efficiency?

The cheapest and easiest way to make headway against our current dependence on fossil fuels is to reduce the demand for energy, which can be done in two basic ways: (1) doing without some of the comforts we've become accustomed to, or (2) improving the efficiency of our energy-using devices. I have friends and neighbors who have done remarkably well at the first strategy through such techniques as walking or biking almost everywhere they go, hanging out clothes to dry instead of using a dryer, and turning their thermostats way down in winter. But while such dedication is admirable, it can be a tough sell to many other people. Indeed, while I like to think that I'm doing my part to help solve the problem, I still often drive even when I have other options, and I use a clothes dryer and keep our home at a comfortable temperature.

The realist in me therefore says that if we want to make big strides in reducing demand, we need to focus our attention on the efficiency side. This is much more possible than you might guess, because there is a great deal of waste in our current energy usage. For example, incandescent light bulbs convert less than 5% of the energy they consume into light, with the rest wasted (primarily as heat). Similar waste is found in almost every other device we use, as well as in our electrical power grid[2] and in the fuel use of cars and airplanes. Therefore, if we can improve efficiency by reducing energy waste, we can maintain (or improve) our lifestyles even while significantly lowering our energy usage. To quote the definition given by physicist and environmental scientist Amory Lovins, energy efficiency means that we can "do the same or more with less."

---

2 The "grid" basically refers to the entire system through which electricity is distributed. In other words, it is the network of power stations and power lines that moves electricity from the places where it is generated to the places where it is used.

As an example, consider the energy used by buildings (both residential and commercial), which goes primarily into heating, cooling, lighting, and other electrical appliances. We can begin to reduce demand simply by installing better insulation and windows, which reduce the energy needed for heating and cooling, and by taking as much advantage as possible of natural light. We can go further by replacing old incandescent light bulbs with newer LED bulbs, which are about three to four times as efficient,[3] and by similarly upgrading other appliances with more efficient ones.

Efficiency gains are also well within reach in other sectors of our economy, including transportation. For example, automakers already have cars on the market that get more than double the gas mileage of the average car driven in America, and doubling gas mileage means we need only half as much fuel to drive the same number of miles. Electric cars might do even better; electric motors are generally more efficient than gasoline engines, so if the power plant supplying the electricity is also efficient (or if the electricity comes from rooftop solar panels), then the total energy required per mile of driving can be reduced substantially. As an example in air transportation, consider Boeing's new Dreamliner aircraft, which typically use some 20% to 25% less fuel per passenger mile than the aircraft they replace.

Overall, improved energy efficiency would seem to be a classic "no-brainer": It allows us to obtain the same energy benefits that we do now while saving us money. Moreover, because it means less total energy use, it can help reduce the future amount of global warming even if we continue to use fossil fuels. Still, as we've discussed, I don't believe that reducing the use of fossil fuels is enough; we need to stop their use entirely. Energy efficiency alone cannot do that, especially when we consider the growing demand for energy in developing nations. With that in mind, let's turn our attention to energy sources that could take the place of fossil fuels.

Could we replace fossil fuels with renewable energy?

The most fashionable alternatives to fossil fuels are renewable energy sources, such as wind, solar, geothermal, hydroelectric, and biofuels. While none of these are perfect — for example, toxic chemicals are used in solar panel production, wind turbines can kill birds, and dams for hydroelectric power can damage river ecosystems — they have the advantage of producing energy without the release of greenhouse gases. The chief debates about renewables therefore focus on

---

3 Future technologies may make lighting even more efficient. For example, in January 2016, researchers at MIT reported on a technique that combines incandescent light bulbs with nanotechnology in a way that, at least in principle, may allow the development of light bulbs that will be at least twice as efficient as LEDs.

how much energy they can realistically provide and whether they are cost-effective.

The easiest way to think about the energy potential of renewables is by considering current total world power consumption, which is about 15 terawatts.[4] The wind carries more than 10 times this much power around the globe, and accessible wind sources are estimated to be enough to supply about 20 terawatts. In principle, then, wind alone could provide for all our current power needs. The potential of solar is far greater: The total amount of solar power reaching Earth is *more than 20,000 times* current world power usage. However, there is significant debate over whether current technology is up to the practical challenges of replacing all of our current energy with renewables. One hurdle lies in the fact that most renewables are intermittent — for example, solar works only when the Sun is shining, and wind only when the wind is blowing — and our current electricity grid isn't well equipped to handle intermittent power. New battery (or other energy storage) technologies may solve this problem, as might changes to the grid, but no one really knows for certain.

With regard to cost-effectiveness, there's great debate about renewables at present. However, as we'll discuss shortly, my personal opinion is that this is a false debate, because the true costs of fossil fuels are actually far higher than we pay. I therefore believe that renewables are already cost-effective, so the only question is whether we can tap enough of their potential to completely end our current dependence on fossil fuels. The answer may well be yes, but until we know for sure, we should keep exploring other existing technologies. This brings us to the topic of nuclear power.

Should we build new nuclear power plants?

Before I give you my answer to the above question, let's consider the major issues with nuclear power. I'll start with the positive side. Nuclear power does not release any greenhouse gases, and it already supplies a significant fraction of the world's energy. In some countries (most notably France) it is the dominant source of electricity. Based on this experience, there's little doubt that nuclear power has great potential as a replacement for fossil fuels, particularly when combined with efficiency and renewables. There's some question as to cost-effectiveness, but as I noted above for renewables, I believe that if we consider

---

4 You can follow this discussion even without a full understanding of the units, but in case you are interested: A *terawatt* is 1 trillion watts, and a watt is a unit of *power*, which means the amount of energy used per second. More specifically, 1 watt of power represents usage of 1 joule of energy per second, so 15 terawatts represents 15 trillion joules of energy per second. This can be more meaningful if you convert it to other energy units. For example, if we obtained all this energy from gasoline, then worldwide we would be burning about 115,000 gallons (435,000 liters) of gasoline *every second*.

4

The Solution

the full costs of energy, nuclear power will also prove to be cheaper than fossil fuels.

Of course, nuclear power also has some well-known drawbacks, including the danger of accidents such as those that have occurred at Three Mile Island, Chernobyl, and Fukushima; the problem of nuclear waste disposal; and the threat that terrorists might obtain radioactive nuclear fuel. The key question for nuclear power, then, is whether these drawbacks can be overcome. The answer is not yet fully known, but there is some cause for optimism.

Let's start with safety. There's no possible way to prevent all accidents, so even with the best safety precautions, we must consider the level of danger posed when some type of accident occurs. Today, the primary danger arises from the fact that existing nuclear power plants use what is called *active* cooling to prevent their nuclear fuel from overheating. This means that if an accident causes a problem in which the cooling system fails, it can lead to a dangerous meltdown and release of radioactive materials. However, new reactor designs have been developed to use *passive* cooling (for example, mixing the fuel with salt so that it expands if the temperature rises, thereby slowing the reactions), in which the power plant would automatically turn itself off if a similar accident occurred. These and other engineering improvements in principle offer the potential to make new nuclear power plants far safer than those built in the past.[5]

The issue with nuclear waste is that it can remain dangerously radioactive for tens of thousands of years, which means it can endanger future generations unless we find a way to keep it isolated enough that no one will ever come across it by accident. This has proved to be a significant challenge, but there are two key "buts." First, some of the new reactor designs that address the safety issue can also significantly address the nuclear waste issue, because they can "reprocess" existing nuclear waste into less dangerous forms. Indeed, many nuclear proponents are confident that with proper design, the nuclear waste problem could be almost completely eliminated. Second, even without a perfect solution, nuclear waste will be dangerous only in the local regions where it is stored. While that is far from ideal, it still seems better than global warming, which threatens the entire planet.

Perhaps the most difficult of the three major issues is the threat from terrorism, since expanded use of nuclear power would mean more radioactive material being transported around the world. Still, technology may offer solutions. For example, the technologies that reprocess waste also leave the remaining materials much less radioactive and therefore unsuitable for building bombs (including "dirty bombs").

---

5   For more detail on some of these issues, I recommend watching the documentary *Pandora's Promise* (2013).

Beyond that, while it will be difficult to protect power plants and their fuel supply, we successfully protect other critical sites, suggesting that it should be possible.

Many thoughtful people have considered all these issues and still come down on opposite sides of the question of whether we should build more nuclear power plants. I myself have vacillated over the years. However, at present, I've come down on the side of "yes," because I'm unconvinced that the combination of efficiency and renewables can end our dependence on fossil fuels soon enough to prevent severe consequences from global warming. In particular, because renewables require not just new power sources (e.g., wind and solar farms) but also changes to the grid and power management techniques, I believe it would take decades to make a full conversion to renewables even under the best of circumstances.

In contrast, I believe that if we were willing to invest sufficient effort and money, we could rapidly replace existing fossil fuel power plants with nuclear power plants that could feed power directly into the existing grid. How rapidly is a matter of great debate, but to me it just comes down to how important we consider it to be. For example, looking back at how rapidly our nation mobilized industry for World War II, it seems to me that it should be well within the realm of possibility to start an effort today that would lead to complete replacement of fossil fuel plants with a combination of nuclear and renewables within about a decade.

Everyone should consider the question of nuclear power carefully for themselves. But if it were up to me, I'd be going full bore into nuclear power, starting immediately, even while we continue to develop renewable energy sources and improve energy efficiency.

So bottom line, do we have the necessary technology to end our fossil fuel dependence?

Based on the discussions above, I think it is clear that we already have the technology to end our fossil fuel dependence. Improved energy efficiency can take us a good deal of the way there, and some combination of renewables and nuclear can take us the rest of the way. I won't claim that it would be easy, but it's certainly doable.

One remaining question concerns whether it's doable globally or only for the United States and other developed nations, since the kinds of solutions we've been discussing are currently quite expensive. Here, again, you'll hear great debate, but as I'll explain in more detail shortly, I'm a big believer in the power of market forces. I therefore believe that if the United States made the necessary investments for the transition away from fossil fuels, the markets would cause the prices of these solutions to fall so much that they would rapidly displace fossil fuels as the cheapest available energy technologies. In that case, the rest of the world would very likely adopt them. Better yet, at least from a U.S. point of view, the rest of the world would then be buying these technologies from U.S. companies, thereby strengthening our nation's economy while benefiting the globe.

Q Why haven't you suggested "clean coal," in which we sequester the carbon dioxide?

Proponents of so-called clean coal suggest that we might continue burning coal for fuel if we can find a way to "sequester" the carbon dioxide and other pollutants, generally by injecting them back into the ground (where they originally came from). Technologically, this may well prove possible, though the verdict on its practicality is not yet clear. However, I have a major concern with this approach due to traditional economics: It seems almost inevitable that sequestering emissions would cost more than allowing them to escape into the atmosphere, and that would give coal plant operators an incentive to cheat and thereby save money. In the developed world, we might be able to prevent such cheating through strict enforcement. But achieving that level of enforcement would likely be very difficult in less developed countries, including India and China. Personally, I therefore find it much better to consider solutions that get us away from fossil fuels altogether. That said, I support continued research into sequestration technology, as it may yet surprise me and prove to be viable.

Q What about fracking and using natural gas as a "bridge fuel"?

The major argument in favor of natural gas is that it releases *less* carbon dioxide per unit of energy generated than oil and coal. In that sense, replacing oil and coal with natural gas is a good thing, particularly if combined with greater energy efficiency that reduces our total energy usage. But there are two main counterarguments. First, natural gas consists primarily of methane, which is a more potent greenhouse gas than carbon dioxide, and studies show that leaks of methane from fracking and other natural gas production can offset much or all of the advantage that would otherwise be associated with this fuel.[6] In other words, unless we find a way to minimize these leaks, natural gas is not always better than oil and coal. Second, a true solution to global warming requires not just reducing but actually stopping the release of carbon dioxide into the atmosphere, and the burning of natural gas still releases carbon dioxide (just not as much per unit of energy as coal or oil). So while natural gas may slow the rate at which the problem becomes worse, it doesn't offer a full solution.

I find these counterarguments compelling, and therefore prefer that we focus our efforts on technologies (such as nuclear and renewables) that could completely end the release of greenhouse gases. It's also worth noting that utilities would have to operate new natural gas power plants for decades to recoup the large investments needed to build them, so the building of such power plants might actually prolong the time until we implement a real solution.

---

6 Just to clarify any potential confusion: When natural gas is burned, methane combines with oxygen to form carbon dioxide. As a result, carbon dioxide is the only greenhouse gas released into the atmosphere by the *burning* of natural gas. The concern I'm addressing here is with methane that leaks directly into the atmosphere before it has a chance to be burned.

Future Energy Technologies

In addition to the existing technologies that likely could already solve the problem of global warming, researchers and entrepreneurs around the world are investigating many other technologies that could in principle do even more. Just to give you a sense of the tremendous potential, I'll introduce you to three of my personal favorites among the many new ideas: fusion, solar energy from space, and microbe-based biofuels.

Why do you think it's worth looking into fusion?

All current nuclear power plants are based on nuclear *fission*, in which atoms of heavy elements such as uranium, plutonium, or thorium are split apart. In contrast, the Sun and stars — and thermonuclear bombs ("H bombs") — rely on nuclear *fusion*, in which hydrogen atoms are fused to make helium. Because the fuel source for fusion is hydrogen, we can get it from water (since water is H_2O). Because its product is non-toxic (and very useful) helium, there's no inherent problem of nuclear waste. Indeed, while details of the engineering would create at least some radioactive material, fusion is probably about the safest, cleanest, and most abundant energy source we know of.

To see how abundant, try the multiple choice question in figure 4.2. If you're like most students I've worked with, you'll probably take a guess from among choices A through D, since E sounds like an implausible throwaway. But E is the correct answer. Think about this fact. If we had the technological capability for fusion and you were willing to let us use your kitchen sink (and leave the faucet water flowing), then we could stop drilling for oil, stop digging for coal, dismantle all the dams on our rivers, take down all the wind turbines, and even turn off all the currently operating nuclear power plants. We'd be able to supply all the energy needed for the entire United States through the fusion of hydrogen extracted from the water flow of your kitchen faucet.[7]

Given such incredible potential, you might wonder what's stopping us, and the answer is simple: Despite decades of effort, no one has yet figured out how to tap nuclear fusion for commercial power. But people are working on it, and the work might well go a lot faster if we devoted more resources to it. There's no guarantee of success, but if it were up to me, I'd make sure we were putting a "Manhattan Project" level of effort

---

7 The calculation that leads to this answer assumes that we could fuse *all* of the hydrogen in the water, just as the Sun fuses hydrogen. In reality, nuclear fusion research generally uses the isotope of hydrogen known as deuterium, which represents only about 1 in 6,400 hydrogen atoms. Therefore, instead of the flow from a kitchen faucet, you'd need the flow from a small creek. Fusion might be even easier to achieve using helium-3 as fuel (ordinary helium is helium-4). Although helium-3 is virtually nonexistent on Earth, there is plenty of it on the Moon, which is one of many reasons why a return to the Moon might provide great economic benefits.

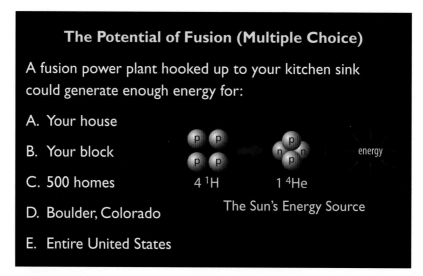

The Potential of Fusion (Multiple Choice)

A fusion power plant hooked up to your kitchen sink could generate enough energy for:

A. Your house

B. Your block

C. 500 homes

D. Boulder, Colorado

E. Entire United States

$4\,^1H$ $1\,^4He$

The Sun's Energy Source

energy

Figure 4.2 Try this multiple choice question about the potential of nuclear fusion, if we can figure out how to tap it as a commercial energy source.

into the development of fusion technology (though being careful that this was not at the expense of efforts to implement the current technologies we've already discussed). Because if we succeed, we'll not only solve global warming, but will also have the ability to generate more energy than we can even imagine what to do with today.

What do you mean by solar energy from space?

As we've discussed, one of the major drawbacks to solar energy on Earth is the fact that it is intermittent, working only in the daytime when it is not cloudy. But it's never cloudy in space, and if you put solar panels in a high enough orbit, it is never nighttime either. So another idea for solving our energy and global warming problems is to launch solar panels into high Earth orbit (figure 4.3), where they would absorb sunlight and beam the energy down to collecting stations on Earth.

It would obviously be expensive to launch solar panels into space, but perhaps not as expensive as you might guess. The "panels" for use in space might well be no thicker than a thin film of plastic wrap, so enormous panels could potentially be unfurled from lightweight spools that could be launched by existing rockets. Moreover, once launched into orbit, these panels would likely provide energy for decades (or more) without the need for replacement. Overall, the total launch costs for enough panels to meet all global energy needs for decades might well be less than we spend (globally) on energy in a single year at present.

Figure 4.3 This painting imagines an astronaut working on solar panels in space, which are used to beam solar energy down to Earth. Painting by Roberta Collier-Morales from *The Wizard Who Saved the World.*

The greater challenges to using solar energy from space probably lie in the technology for transmitting the energy to Earth, in building the collecting stations for that energy, and in tying those stations into a power grid that could distribute the energy around the world. But none of these challenges appear to be insurmountable, and proponents of solar energy from space argue that we could start implementing it now if we were willing to make a concerted effort on it. To learn more about this promising technology, a good starting point is this Web page from the U.S. Department of Energy: energy.gov/articles/space-based-solar-power. You'll also turn up more information if you search on "space-based solar power."

Q What are microbe-based biofuels?

Biofuels are fuels made from plants or other living organisms. The best known biofuel has been ethanol from corn, which turns out to be very inefficient because corn agriculture is so energy intensive, and because it diverts land that can be used to grow food (which we also need!) to energy production. But microbial biofuels, such as those made from algae or from bioengineered organisms, offer far greater potential.

Microbe-based biofuels have the potential to stop global warming because even though they release carbon dioxide when burned, the microbes absorb it when they grow. This means that if you don't use too much energy in making the biofuels, they can essentially be "carbon neutral," meaning that the microbial growth absorbs exactly as much carbon dioxide as the fuels release when burned. In fact, some researchers are developing biofuels that might actually give us a net *reduction* in atmospheric carbon dioxide (by taking up carbon dioxide and converting some of it into minerals or other forms that would permanently remove it from the atmosphere), meaning they could potentially help reverse the damage that has already been done. It's also worth noting that biofuels are much easier to use as fuels for airplanes and rockets than electricity generated by centralized power plants, so they may well prove to be of great importance even if we successfully implement other solutions such as fusion or solar energy from space.

Perhaps the best news about microbe-based biofuels is that they already exist. For example, algae-based biofuels have been successfully tested in commercial airplanes and in Navy ships. Their overall potential is still a matter of debate, but there's no question that they could make an important contribution to our future energy supply.

Keep in mind that biofuels, solar energy from space, and fusion are just three of my personal favorite ideas for the future. People are working on many others, which brings us back to the key point: *If we put our minds to it, we can develop technologies that won't just solve our current problems, but will make the future better and brighter for the entire human race.*

What about geoengineering?

Imagine that, a couple of decades from now, global warming is rapidly worsening and no new technologies have been successfully implemented to stop it. What then? This scary possibility has led some people to ponder *geoengineering* schemes in which we would deliberately try to alter Earth's climate in ways that might counter the planetwide effects of global warming. For example, some people have proposed seeding the atmosphere with aerosols that would reflect sunlight back to space, or even deploying giant sunshades in space. While I support continued research into these ideas, I have left them out of my discussion of solutions to global warming for a simple reason: With one exception that I'll explain below, I don't think we could ever be confident that any of these "cures" for global warming would be better than the disease.

The reason for my pessimism about most geoengineering schemes is that they would not by themselves reduce the amount of carbon dioxide we are releasing into the atmosphere. They therefore suffer from at least three major drawbacks. First, because they allow the carbon dioxide concentration to continue to increase, they do nothing at all about the problem of ocean acidification. As we've discussed, this problem is probably at least as serious as any of the other consequences of global warming, so a "solution" that leaves it unaddressed does not seem to be a real solution. Second, most geoengineering schemes require active maintenance; for example, the aerosol idea requires continually putting more aerosols in the atmosphere to replace those that rain out, and even the sunshades in space would likely need occasional orbital adjustments. If the maintenance ever failed — whether now or centuries from now — global warming would immediately resume, and if we'd continued adding carbon dioxide in the interim, it would be far worse than it is today. Third, these types of geoengineering introduce global climate factors that do not exist naturally and therefore are difficult to account for in models. As a result, we don't have any good way to predict the full consequences of these schemes, so even if they successfully stopped the rise in Earth's average temperature, we could not be confident that they wouldn't create regional climate disruption.

All that said, there is one important exception, and that is geoengineering intended to actually *remove* carbon dioxide from the atmosphere. There are a number of technologies under development that might actually be able to do this (including the biofuels we've already discussed), essentially by absorbing carbon dioxide from the atmosphere and then incorporating it into rock or some other solid form that can be safely stored away. If any of these technologies could remove atmospheric carbon dioxide faster than we add it, then they might not only offer a real solution to the problem of global warming but also be able to *reverse* some of the damage that has already been done. For this reason, I believe it is critically important that we work on these technologies and implement them if they become available.[8]

Q If future geoengineering technologies may be able to reverse global warming, can't we just wait for those instead of dealing with the problem now?

Consider a medical analogy. Imagine that you have a potentially fatal disease for which scientists are working on a cure, but you also know that you can slow its progress if you quit smoking and improve your diet. Would you

---

8 Of course, even these technologies would require great care. For example, they would have to be easily controlled, so that there would be no risk that we might inadvertently remove so much carbon dioxide that we'd plunge our planet into an ice age.

continue smoking and eating poorly in hopes that the cure comes before you die? I certainly hope not, and in the same way, it makes no sense to continue making the problem of global warming worse while we search for ways to reverse it. Moreover, the sooner we stop worsening the problem, the more likely that any future "cure" will be successful. So, as medical workers learn, first stop the bleeding.

The Obstacle to a Solution

I hope that I've convinced you that solutions to the problems of global warming already exist and will become even better in the future. So you might wonder, what's stopping us?

The obvious answer is pricing. For individuals (or individual companies), it generally remains cheaper to buy energy created from fossil fuels than to implement any of these other energy solutions. This is especially true as I write in early 2016, when oil prices have plunged to their lowest levels in many years. But here's the thing: It's *not cheaper for society*, because there are a great many costs associated with fossil fuels that are very real but that are incurred by society as a whole rather than by the people who use the energy.

Economists refer to these costs as "externalities," meaning costs that are not incorporated into the actual ("internal") price we pay for energy. But I prefer to think of them in a different and admittedly more provocative way: I'd argue that our current energy economy is essentially a form of *socialism*. I see little difference between the way we socialize many of the true costs of energy and the way some countries socialize medical or other costs. To help you understand why, let's go to our next question.

Are there socialized energy costs that almost everyone agrees on?

By a *socialized cost* of energy, I mean any cost that is real but that is borne by society as a whole rather than by individual energy users. There are many such costs, some of which are easier to quantify — and hence subject to less dispute — than others. In particular, there are three categories of socialized costs that almost everyone agrees on to at least some extent: (1) health costs associated with pollution from fossil fuels; (2) military costs of protecting the fossil fuel supply; and (3) the costs of direct subsidies and tax write-offs for fossil fuel companies.

Let's start with the health costs. Besides emitting carbon dioxide when burned, fossil fuels release many other pollutants into the air and water, and these have real costs for human health that are borne by society as a whole through taxes, medical insurance premiums, and other shared health care costs. For example, in 2010 the U.S. National

Figure 4.4 These two photos were taken from the same location in Beijing on a clear day and a smoggy day. Beijing air is frequently so polluted that there is substantial health risk just in being outside, and pollution may well be contributing to substantially reduced life expectancy in Beijing and many other highly polluted cities in the developing world. Credit: Bobak Ha'Eri, Wikimedia Commons.

Academy of Sciences estimated the direct health costs of air pollution to be about $120 billion per year. Note that by *direct* costs, they mean only those that are quantifiable in terms of actual medical bills or loss of life. On top of this, there are indirect costs for lost workdays and reduced worker productivity when someone is ill, and the study did not include the costs of water pollution. I don't think it is any stretch to assume the total health costs associated with pollution exceed $200 billion per year in the United States. And if you have paid attention to the extreme pollution that now occurs regularly in other parts of the world (figure 4.4), then you'll realize that the global costs are probably far higher.

The military costs are a little more difficult to pin down, because while there's no doubt that the United States spends a lot of money on military efforts that protect the international oil supply, these same efforts also serve other purposes. For example, the same Navy vessels that protect the shipping lanes used for Middle Eastern oil also serve in the fight against international terrorism. Nevertheless, studies by the U.S. Department of Defense and others have tried to quantify the cost of protecting the international oil supply; while precise conclusions vary, these reports seem to converge around an estimated cost of about $100 billion per year. Note that this is only the baseline cost (protecting shipping lanes, etc.) and does *not* include costs (both human and economic) of wars that have occurred in oil-rich regions or of interventions against enemies that rely on oil revenues to support their efforts against us.

The third category of generally agreed upon costs takes two forms: direct subsidies to fossil fuel companies for fossil fuel exploration and production, and tax write-offs for their own expenditures on these activities. Both of these have the same net effect of costing taxpayers real money, and in the United States they are estimated to add up to at least about $20 billion per year.

Notice that the above costs already total more than $300 billion per year for the United States alone. As an example of what this would mean if we asked users to pay these socialized costs, let's suppose we decided to build them into the price of gasoline through a gasoline tax. Drivers in the United States use a total of around 150 billion gallons of gasoline per year, so charging users for the socialized $300 billion would require a new gasoline tax of about *$2 per gallon*.[9] Most people find this amount to be surprisingly high, but as we've discussed, this would only stop the socialization of costs that almost everyone agrees are very real.

What additional energy costs are currently socialized?

Many other socialized costs are clearly traceable to our use of fossil fuels, though they are more difficult to quantify. For example:

- Costs of environmental damage from such causes as oil spills, strip mines, and water contamination. These can be difficult to quantify because they depend on what value we assign to the damaged environment, but economists who have made the attempt suggest that they represent at least tens of billions — and possibly hundreds of billions — of dollars per year in the United States, and much more globally.
- Costs to the national economy of purchasing imported energy. These are also difficult to quantify but, for example, a U.S. Department of Energy study ("Costs of Oil Dependence: A 2000 Update," by D. L. Greene and N. I. Tishchishyna) concluded that our dependence on imported oil had cost the United States about $7 trillion in lost wealth during the period 1970 to 2000, which translates to about $250 billion per year.
- Human costs of terrorism and totalitarianism that have been fueled by oil revenue. There's no doubt that over recent decades, the enemies of democracy and freedom have been funded largely by revenue from oil, and they have used this revenue to spread terrorism, intolerance, and hatred around the world. It's difficult to put a price on this destruction, but if we tried, I suspect it would greatly exceed the price of everything else we've discussed so far.
- Costs associated with global warming. While these may be relatively limited to date (though still many billions of dollars), the future consequences could easily cause these costs to rise above all the others combined.

---

9 By charging all of the $300 billion in direct costs to gasoline, I'm assuming that we essentially give a pass to fossil fuels used in power plants and industry. As we'll discuss later, a better approach is probably a carbon tax rather than a gasoline tax, so that all fossil fuel use is treated similarly in terms of the economy.

The uncertainties in pinning down these costs haven't stopped people from trying to estimate them, and even conservative economists have often come up with astonishingly high total values for all the socialized costs of fossil fuels. For example, in 2006, energy analyst Milton Copulos — who worked for both the Reagan administration and the conservative Heritage Foundation — concluded that the "hidden" costs of oil alone (that is, not even including costs associated with coal and natural gas) totaled $780 billion per year in the United States. A 2015 study by the International Monetary Fund (IMF) calculated the total *global* value of subsidies for fossil fuels at more than $5 trillion per year, of which more than $4 trillion represented costs before global warming was even taken into account.[10]

The bottom line is that no matter how you look at it, the full costs of fossil fuels are not incorporated in their current market prices. Instead these costs are socialized, meaning they are borne by taxpayers and society as a whole. In fact, they are even socialized across generations, because some real costs, such as those associated with global warming, will be borne primarily by our children and grandchildren (and subsequent generations). That is why I say we have a socialist energy economy today.

Q Are you saying that politicians who favor the status quo of energy pricing are secretly socialists?

I try not to engage in name calling, but I'm always surprised that so few people have paid attention to the clearly socialized costs that we have discussed, and by the associated irony that people who call themselves free-market conservatives have generally been the ones most resistant to any change in the energy status quo. I therefore suspect that these politicians simply have not studied the issue enough, and I hope that once they understand the facts of the matter, they will decide to apply their free-market principles to energy in the same way they apply them to other aspects of our economy. I also hope they will take to heart the definition of conservatism offered by Ronald Reagan in the quote that opens chapter 5 (page 95).

The Clear Pathway to the Future

There may be plenty of room for debate over the particular numbers that I've given you for the socialized costs of fossil fuels, but there's no doubt that the true costs of these fuels are substantially higher than what individuals and companies currently pay for them. My numbers

---

10 The report from Milton Copulos is described in his 2006 testimony to Congress, reproduced at www.evworld.com/article.cfm?storyid=1003. The IMF report is summarized at blog-imfdirect.imf.org/2015/05/18/act-local-solve-global-the-5-3-trillion-energy-subsidy-problem.

suggest that they are higher by the equivalent of at least a couple of dollars per gallon of gasoline, and it's quite likely that the true costs are *several times* the current market prices. These ideas lead to what I believe to be the simple "win-win" solution to the problems we face with respect to energy, pollution, and global warming.

How can we institute a true free market for energy?

Imagine that instead of socializing the costs of fossil fuels, we charged these costs directly to the individuals and companies that use the fuels. We would then be able to make energy decisions through a fair comparison between the true prices of fossil fuels and other energy sources. I'm a big believer in the power of free markets, and I believe that if such a free market for energy existed, we'd find that the true prices of wind, solar, and nuclear are all already substantially lower than the true prices of fossil fuels. In that case, a free market for energy could rapidly lead to a solution to global warming, because everyone would have strong financial incentives to replace fossil fuel power with power from less expensive alternative sources. A free market would also encourage entrepreneurs and businesses to invest more in researching new technologies, because the potential payoff would be much more lucrative than it is with our current system, which keeps fossil fuel prices artificially low.

The seemingly obvious way to institute a true free market for energy is to build the currently socialized costs into the prices paid in the marketplace. Economists across the political spectrum agree that there's one surefire way to do this: Institute a *carbon tax* that accounts for the socialized costs, so that market prices (with the carbon tax included) will reflect the true costs of fossil fuels. Indeed, this solution is so widely accepted among economists of all political stripes that you'll rarely see any debate about its general validity. Instead, the debate focuses on two points relating to the implementation of a carbon tax: (1) how high it should be, and (2) what to do with the revenue it generates.

With regard to the first point, in principle the carbon tax should be high enough to account for *all* the socialized costs of fossil fuels. In practice, these costs are almost certainly so high that, at least in the short term, we could not institute the appropriate tax without great risk to the economy. But this is an easy problem to deal with: We simply introduce the carbon tax gradually, so that individuals and companies have time to adapt as it rises.

To the second point, the most common answer is that the revenue should be returned to those who bore the socialized costs, which means taxpayers and society at large. Broadly speaking, there are three possible ways of doing this: (1) We can lower other tax rates to offset the revenue coming in from the carbon tax; (2) We can provide "dividends" to all members of society from the revenue; and (3) We can fund government projects or services intended to benefit society. As you might guess, people of different political persuasions come to different conclusions about how those three possibilities should be weighted. For example, conservatives tend to lean toward lowering other tax rates, while liberals tend to favor dividends and/or increased government spending.

Personally, I'm much less concerned with what we do with the revenue than in making sure we institute a carbon tax so that the free market can take care of the critical problem of global warming. That said, if it were up to me, I'd institute what economists call a "revenue-neutral" carbon tax, meaning one in which all the incoming revenue would be returned to the public through some combination of lower tax rates or dividends.[11] This is not because I'm necessarily opposed to more government spending, but because I think the question of appropriate spending levels should be kept separate from the question of how we solve the problem of global warming. Between lower tax rates and dividends, I favor a combination that is weighted toward dividends distributed more or less equally to everyone. The reason is that a carbon tax will necessarily have a greater impact on the poor than on the wealthy, and my personal sensibilities favor something that helps the poor offset this impact. Tax rate reductions offer little benefit to those already paying very little in taxes, but dividends represent cash in hand that can make one's life a little easier.

To summarize, a carbon tax offers the simplest and best opportunity for creating a true free market in energy. Perhaps I'm overly optimistic about the power of free markets, but I believe that if we successfully created a free market for energy, we'd rapidly find solutions to the problem of global warming, all while strengthening our overall economy and improving our lives at the same time. For that reason, regardless of your political persuasion, I hope you will join in the growing movement for a carbon tax that will help end the current socialism of our energy economy.

Q What about "cap and trade" instead of a carbon tax?

"Cap and trade" is an alternative approach that, like a carbon tax, is intended to encourage market forces that would lead to more investments in alternative technologies that don't emit greenhouse gases. You can read more about this approach by searching on "carbon trading," but in brief it works like this: The government places a legal limit ("cap") on the total carbon dioxide emissions that are to be allowed, then sells, auctions, or initially gives away permits that allow companies to emit portions of this

---

11 One example of this type of approach is ballot initiative 732 that will be voted on in Washington State in November, 2016.

total. Companies can then buy and sell ("trade") these permits so that, for example, a company emitting more carbon dioxide than it is allowed can buy additional permits from a company emitting less than it is allowed. In principle, this encourages companies to reduce emissions since they can profit by selling their permits to other companies that have been less innovative. If the cap is lowered over time, the total emissions should go down.

Cap and trade systems have been used successfully for other pollutants; a notable example is a system in the United States that has led to dramatic reductions in the emissions that cause acid rain. For greenhouse gas emissions, cap and trade has already been implemented in the European Union and several other nations, and in the United States by California and the states involved in the Regional Greenhouse Gas Initiative, though there is debate about how successful these efforts have been to date. A bill to institute a cap and trade system was passed by the U.S. House of Representatives in 2009 but died in the Senate.

It's certainly possible that cap and trade could successfully reduce greenhouse gas emissions, but my own opinion is that it is inferior to a straight carbon tax for at least two major reasons: (1) It is much more complex than a straight tax, since it requires the creation and maintenance of markets for the permits, and (2) unlike a carbon tax, which makes it easy for the public to see the true cost of the fuels they are purchasing, the behind-the-scenes trading of the cap and trade system will leave much of the public unsure why different energy sources cost what they do. Given these drawbacks, I see no reason for the added complexity of cap and trade when a much simpler carbon tax is likely to be just as effective, if not more so. That said, if it turns out politically to be easier to do cap and trade, it's certainly far better than nothing.

Q What about other governmental approaches to dealing with global warming, such as subsidies for solar and wind, regulations supporting mileage standards for vehicles, and so on?

In the past, I have supported all of these things, because they represent small steps toward dealing with the problem of global warming. But if we were to institute a real carbon tax as discussed above, I believe other subsidies and regulations would become unnecessary, because the free market for energy would take care of the problem by itself. That is why I believe this solution transcends politics: Whether or not you favor government regulation in general, and whether or not you favor an unfettered free market in general, in this particular case the market economics are sufficiently clear to make the case for a carbon tax one that almost everyone can agree on.

5 A Letter to Your Grandchildren

What is a conservative after all but one who conserves, one who is committed to protecting and holding close the things by which we live . . . our countryside, our rivers and mountains, our plains and meadows and forests. This is our patrimony. This is what we leave to our children. And our great moral responsibility is to leave it to them either as we found it or better than we found it.

— President Ronald Reagan, June 19, 1984 (remarks at the dedication of the National Geographic Society headquarters building)

My greatest worry is the consequence of not stepping up, of someday having my own kids ask me, "When the stakes were so high, why didn't you do all you could do, why didn't you stand and fight for my future?"

— Congressman Paul Ryan (on becoming Speaker),[1] Oct. 20, 2015

In this short book, I have only touched on the many details that lie behind the issues of global warming, its consequences, and its potential solutions. After all, if we consider just the science alone, thousands of scientists around the world have devoted their lives to studying the climate and how global warming will affect it. No single person could ever become an expert in every subtopic of the research. Nevertheless, I hope that by now I have convinced you that, despite what you may hear in the media and from politicians who choose to remain ignorant, the basic facts are simple and clear. Let's summarize them one more time, in a slightly different way than we have before:

- The laws of physics tell us that, because we are adding greenhouse gases to the atmosphere, we should expect global warming to be occurring.

---

1 Ryan made this statement in the context of his decision to accept the nomination for Speaker of the House, but I've used it here because it expresses a general sentiment that applies in many other contexts as well.

- Data collected over many decades by thousands of scientists support the conclusion that the warming is occurring as expected and that this warming will be detrimental to our future.
- While the problem of global warming will not be easy to solve, we have the technological capability to implement solutions that will not only prevent us from suffering its worst consequences but will actually lead us to a stronger economy and better lives.

With that in mind, I'll close by suggesting a more personal way of thinking about the role that *you* can play in solving this important problem.

On average, we are all about 50 to 60 years older than our grandchildren are or will be. For example, if you're reading this in 2016 and you are a high school student, your future grandchildren are likely to be in high school some time around the year 2066. If you are in your prime career years today, your grandchildren will be in theirs in 2066. If you are a senior citizen today, your grandchildren will be approaching their own senior citizenship in 2066. So imagine sealing a letter in a time capsule for your grandchildren to open 50 years from now, like the sample one in figure 5.1. What will you say to them about what you did — or did not do — in the face of today's understanding that global warming poses a real and serious threat to their future?

In case it's still not clear, my own way of completing the letter is simple. I will continue to do everything I can to educate people about the simple and basic science that lies behind the issue of global warming; I will do everything in my power to push for a carbon tax like the one discussed in chapter 4, since I believe that is the clearest and simplest path to solving the problems; and I will support only those politicians who acknowledge the reality of global warming[2] and are willing to take serious steps to protect the futures of our children and grandchildren. I hope you will join me in making similar commitments to the future.

---

2 Note that this is *not* an example of "single-issue voting," but rather a case of asking politicians to demonstrate an ability to be intelligent and thoughtful. After all, in the face of the tremendous body of evidence we've discussed in this book, anyone who still refuses to take the issue of global warming seriously would seem undeserving of our trust on national security, the economy, and any other complex issue.

— To be opened in 50 years —

Dear Grandchildren,

As I write this in 20XX, many people are arguing about whether global warming is a real problem and, if so, how serious it will be for you when you are my age. I have examined the evidence, and I have decided to . . . [fill in your decision]

Hope your world is a good one.

Love,

_____
[your name here]

Figure 5.1 On average, 50 years from now your grandchildren will be about as old as you are today. Complete this letter and put it in a time capsule for them to open in 50 years. Then ask yourself, how will they feel about what you did — or did not do — to help secure their future?

Acknowledgments

Although I'm listed as the author of this book, I hardly deserve the credit, since all the research I describe has been done by others. So my first acknowledgment goes to the many scientists around the world who have dedicated their lives to helping us understand global warming in all its details and implications. I'm also deeply indebted to the many scientists who have put simple and clear information out on the Web, especially those behind the Web sites listed on the "To Learn More" page that follows. These sites were invaluable to me in researching the many details included in this book.

The pedagogical approach of this book is one that I've developed with a great deal of help from numerous other people, especially my textbook coauthors Nick Schneider, Megan Donahue, and Mark Voit. Additional thanks to Dr. Schneider for coming up with the four skeptic claims (in slightly different form) that form the backbone of chapter 2.

The book greatly benefited from careful review by many readers. I'll start with Steve Montzka (NOAA), who did the first expert review of the manuscript and gave me many great suggestions for improvements. Additional expert reviewers who provided great assistance include Scott Mandia, William Becker, David Bailey, David Bookbinder, James McKay, Kirsten Meymaris, Greg Meymaris, Shawn Beckman, Yoram Bauman, John Bergman, Minda Berbeco, Glenn Branch, Piers Forster, William Gail, Dave Cleary, and Roger Briggs. Several nonscientist reviewers gave me great feedback as well, including Mark Levy, Helen Zenter, Susan Nedell, Val Wheeler, and my longtime editor Joan Marsh.

For producing the book, I thank Mark Ong and Susan Riley of Side By Side Studios in San Francisco. Mark and Susan not only did all the design work, but also provided many great suggestions on the approach and content. The book Web site (globalwarmingprimer.com) looks great thanks to Web maestro Courtney Faust of Saffron Design in Boulder, Colorado. Finally, I thank my wife, Lisa, and my children, Grant and Brooke, for their ongoing support, inspiration, and insights.

To Learn More

Human history is more and more a race between education and catastrophe.

— H. G. Wells, 1920

There are far too many great resources on global warming for me to list them all in one place, but I'll offer a few of my personal favorites, by category:

Books for adults: If you only read one additional book on global warming, I'd recommend *The Discovery of Global Warming* by Spencer Weart (Harvard University Press, 2008), which details how we've learned what we know simply and clearly. Dr. Weart also has posted an extended version of the book at www.aip.org/history/climate/.

Books for children: I hope you won't mind if I recommend my own, *The Wizard Who Saved the World* (Big Kid Science, 2011). It explains the basic science of global warming while offering children an optimistic and inspirational view of our future and how they can contribute to it. The book was selected for launch to the International Space Station as part of the Story Time From Space program, and was read from orbit by Japanese astronaut Koichi Wakata. A video of his reading is posted at www.StoryTimeFromSpace.com.

Video: For a great general overview of the topic in about 45 minutes, I recommend episode 12 ("The World Set Free") of the updated *Cosmos* series with Neil deGrasse Tyson, available on Netflix and elsewhere. The National Academy of Sciences has produced a video set (26 minutes total) called "Climate Change: Lines of Evidence" that summarizes the evidence for global warming (alturl.com/q3jt3). If you have more time, you can watch the Showtime documentary series *Years of Living Dangerously*, which has nine hour-long episodes.

Web sites:
- climate.nasa.gov: NASA maintains this site focused on global climate change with outstanding summaries and regularly updated news.

Japanese astronaut Koichi Wakata reads *The Wizard Who Saved the World*, a children's book that discusses the science of and solutions to global warming, aboard the International Space Station. A video of the reading is posted on the web site of the Story Time From Space program (www.StoryTimeFromSpace.com).

- skepticalscience.com: For questions beyond those addressed in this book, the Skeptical Science Web site is the place to go for answers. Especially see their list of "climate myths" in the left column.
- climatecentral.org: The Climate Central Web site has a compendium of almost everything you might want to know about global warming and all the latest news.
- www.ipcc.ch: This is the place to find the latest reports from the Intergovernmental Panel on Climate Change, which represent the work of thousands of climate scientists from around the world.
- A few of many other great sites include climatecommunication.org, realclimate.org, c2es.org, yaleclimateconnections.org, and www2.sunysuffolk.edu/mandias/global_warming/.
- For those interested in looking beyond global warming to other important issues that will affect our children and grandchildren, I suggest checking out www.contractwiththefuture.org.

Finally, you can find additional materials, including information about booking me to speak on this topic, at the book web site

www.globalwarmingprimer.com

Index

About the Author

Dr. Jeffrey Bennett holds a B.A. in Biophysics (University of California, San Diego) and an M.S. and Ph.D. in Astrophysics (University of Colorado). His extensive educational experience includes teaching at every level from preschool through graduate school, proposing and helping to develop the *Voyage Scale Model Solar System* on the National Mall in Washington, DC, creating the first broad-based curriculum for courses in quantitative reasoning, and serving two years as a Visiting Senior Scientist at NASA Headquarters, where he helped create numerous programs designed to build stronger links between the research and education communities. He is the lead author of college textbooks in astronomy, astrobiology, mathematics, and statistics that together have sold more than 1.5 million copies, and of numerous critically acclaimed books for the general public, for children, and for educators. He has received numerous awards, including the American Institute of Physics Science Communication Award, and has been honored to have seven of his books launched into space: His astronomy textbook, *The Cosmic Perspective*, flew on the Space Shuttle Atlantis during the final servicing mission to the Hubble Space Telescope in 2009, and his six children's books are all currently aboard the International Space Station, where they have been read aloud by astronauts for the Story Time From Space program. For more details, please see his personal web site, www.jeffreybennett.com.